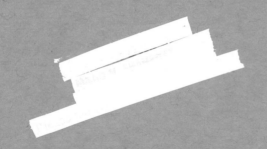

get a grip on
EVOLUTION

Get a Grip on
EVOLUTION

DAVID
BURNIE

TIME®
LIFE
BOOKS

Time-Life Books is a division of Time Life Inc.

TIME LIFE INC.
President and CEO: George Artandi

TIME-LIFE CUSTOM PUBLISHING
Vice President and Publisher   Terry Newell
Vice President of Sales and Marketing Neil Levin
Project Manager   Jennie Halfant
Director of Creative Services   Laura Ciccone McNeill
Director of Acquisitions   Jennifer Pearce
Director of Special Markets   Liz Ziehl

This book was conceived, designed, and produced by The Ivy Press Limited,
2-3 St. Andrews Place, Lewes, East Sussex BN7 1UP, England

Art Director   Peter Bridgewater
Editorial Director   Sophie Collins
Designer   Angela English
Commissioning Editor   Andrew Kirk
Picture Research   Vanessa Fletcher
Illustrations   Andrew Kulman

Reproduction and printing in Hong Kong by
Hong Kong Graphic and Printing Ltd.

TIME-LIFE is a trademark of Time Warner Inc. U.S.A.

LIBRARY OF CONGRESS CATALOGING-IN-PUBLICATION DATA
Burnie. David
    Evolution/David Burnie
        p.   cm.   -- (Get a grip on--)
    Includes index.
    ISBN 0-7370-0036-8 (alk. paper)
    1. Evolution (Biology) I. Title. II. Series.
    QH366.2.B87   1999
    576.8--dc21   98-52940
                                                              CIP

CONTENTS

INTRODUCTION

GET A GRIP ON EVOLUTION

* In the entire history of science, few ideas have generated such passionate debate as evolution. Yet for the majority of scientists today, life and evolution are as inseparable as matter and gravitation. The notion that living things slowly adapt and change as generations succeed each other is so well validated that it is no longer simply a theory, but a <u>PARADIGM</u> that shapes every aspect of the science of life.

living things are capable of change

Evolution and predictability

Like all natural phenomena, evolution is subject to laws that can be verified by experiment. However, the many random elements involved in evolution mean that its future path can never be predicted.

REVEALING CHANGE

* For nearly 150 years the concept of evolution has been associated with one man: the English naturalist **Charles Darwin (1809–82)**. However, Darwin did not "invent" the idea of evolution, nor did he single-handedly spring it onto the world. His achievement was to present the evidence for evolution in a coherent way, showing that living things are capable of change. This evidence was summed up in *The Origin of Species*, a book that sold out the day it went on sale, in 1859, and which has been in print ever since.

✱ Nevertheless, at the time of his death, Darwin's achievement was only partly complete. Although he had persuaded most of his peers of the possibility of evolution, the chief driving force that he proposed—NATURAL SELECTION—attracted only a small band of dedicated adherents. It was only in the early 20th century, with the discovery of the mechanisms of heredity, that Darwin was vindicated.

EVOLUTION'S ENEMIES

✱ Evolution has profound implications for us as a species and for our view of the world around us. During the 19th century, evolution was seen as a threat to the very foundations on which society was built, because it ran counter to the teachings that every living thing—and every person—had its own fixed station in life. During the 20th century, evolution's underlying mechanisms have helped to foster what some see as a bleak view of the natural world, in which living things are the mere playthings of chance.

✱ In the development of evolutionary theory, scientific objectivity has often run up against moral sentiment and religious beliefs. Darwin may have won over the scientific community, but in the wider world, opposition to the idea of adaptation continues today.

KEY WORDS

PARADIGM:
an overall framework of thinking within which scientific theories are constructed and assessed

HEREDITY:
the mechanisms involved in passing characteristics from one generation to another

NATURAL SELECTION:
an environmental process that allows the fittest forms of living things to leave the most descendants

evolution has met with a lot of opposition

could it possibly be...

EXTRAVAGANT LIFE

* To date, biologists have identified and named about 2 million species of living things. Estimates for the total number that actually exist vary from 10 million to a staggering 50 million or more. However, despite all the time and energy that has been devoted to identifying different forms of life, defining exactly what a species is has proved remarkably difficult.

identifying the species

WHAT IS A SPECIES?

* The term SPECIES is derived from a Latin word meaning a kind or class of thing. That sounds simple enough—even though, confusingly, the term "species" can be singular or plural. A more deep-seated problem lies with the idea of species itself. For more than two centuries, biologists have struggled to come up with a simple definition, despite the fact that at first glance the idea seems almost too simple to need defining.

* *The most unambiguous features of species is that they look different: a swallow, for example, is different from a hawk, and a rabbit is different from a rat.* But if you probe a little deeper into the natural world, things rapidly get more complicated. Does the Canadian lynx (also known as the bobcat) belong to the same species as the lynx that lives in Europe? Is the royal penguin a different species to the very similar macaroni penguin? Are there—as some botanists believe—more than 250 different dandelion species in Europe, or are there just a handful of species, with many local variants? *The difficulty with differences like these is that they are susceptible to interpretation: some people see major differences where others do not.*

what sort of penguin am I?

A DETACHED APPROACH

✱ *To avoid these problems, species are now not usually defined solely on the basis of physical characteristics, but also on a biological basis. The key part of this concept is that, as far as breeding is concerned, species keep themselves to themselves: they display what is known as* <u>REPRODUCTIVE ISOLATION</u>. *By breeding only with their own kind, they maintain a pool of characteristics that are found in that species and not in any other.*

Hmmm...

✱ *This feature of species has an important implication. It means that although the Earth teems with life forms, and species interact in countless ways, each one is cut off from every other form of life.*

9

THE MYTH OF PERFECTION

* Living things are often described as being "perfectly adapted" to their way of life. But if this were actually true, any form of change would be bad news—because it would necessarily be a change for the worse. In reality, change continues all the time, because natural selection constantly updates living things, tailoring them to fit their environment.

I'm sure I'm missing a bit

LEFTOVER LETTERS

In *The Origin of Species*, Darwin compared vestigial organs to letters that remain in the spelling of a word, but which are no longer pronounced: although they serve no useful purpose, they hint at how the word has evolved.

whales still have their pelvic girdle

STAYING IN FRONT

* Before the idea of evolution became widely accepted, living things were assumed to be individually designed by the Creator, and therefore perfectly suited to their own way of life. *Today's biologists see the natural world in a very different light. Instead of being in a static state of perfection, every species is constantly adjusting to the world around it.* The result of these adjustments—over many generations—is ADAPTATION.

* Adaptation is a process that can never be complete, because change is built into an organism's environment. The physical environment changes, and the biological environment changes as well, as organisms adapt to the

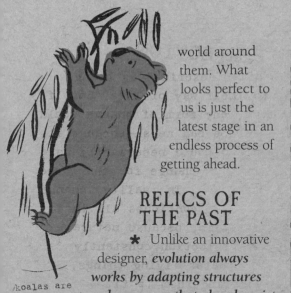

koalas are adapted to eat eucalyptus leaves

world around them. What looks perfect to us is just the latest stage in an endless process of getting ahead.

RELICS OF THE PAST

✱ Unlike an innovative designer, *evolution always works by adapting structures and processes that already exist, rather than starting from scratch. If something loses its adaptive value, it cannot simply be junked—instead, it usually remains present in an increasingly reduced form, for a long period of time. This means that, far from being perfect, living things often have features that no longer serve any useful purpose at all.*

✱ These evolutionary leftovers are common in the animal world. Kiwis, for example, have tiny wings which, at only two inches long, are completely useless for flight. Whales still have the remains of a pelvic girdle, even though they have no back legs, and humans still have the tiny vertebrae that once formed part of a tail. *These built-in "defects" not only undermine the idea of perfection, they also throw light on the way different species have developed.*

KEY WORDS

ADAPTATION: the gradual accumulation of inherited characteristics that give living organisms the best chances of surviving and reproducing; the word can also be used to mean the inherited characteristics themselves

Problem pouch

The human appendix is an evolutionary leftover that probably helped our distant ancestors digest a mainly plant-based diet. This finger-shaped pouch branches off the first part of the large intestine, but its precise position varies from one person to another. It no longer performs any known function and can become dangerous if inflamed—so has turned from an asset into a liability.

human appendix

11

THEMES AND VARIATIONS

* With the exception of identical twins, no two people look exactly alike, which makes it easy for us to recognize each other. In the world of nature, individual differences are sometimes less obvious, but they exist just the same. VARIATIONS are passed on by inheritance, and they form the raw material of evolution.

I'm more attractive...

no two people are exactly alike

THE VALUE OF BEING VARIED

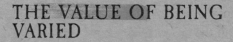

* In many living things, sameness seems to be a fact of life. One housefly looks just like any other, and the same is true of fish in a school, or penguins crowded together in a colony. Plants are sometimes affected by local conditions, but even so, one daisy or poppy looks much the same as the next.

* In fact, these superficial similarities are deceptive, because *within each species there is a vast pool of subtle variation*. Some of these variations are expressed as outward characteristics—such as shape, size, and color—but others affect factors that are much harder to discern, such as growth rates, ability to break down certain nutrients, or resistance to disease.

The small print

In the natural world, as in the financial one, past performance is no guide to what may happen in the future. There is no guarantee that a useful characteristic will remain useful indefinitely, and its spread throughout a species may be checked long before it has had time to become common.

Regardless of whether we notice variations or not, they all play a potential role in the struggle for biological success.

FROM MINORITY TO MAINSTREAM

you'll grow up to look just like me

✱ In the daily lottery of life, living things are constantly put to the test, and some fare better than others. A particular characteristic may reduce an individual's chances of breeding successfully or, alternatively, may increase its chances of leaving offspring. Even if that increase is only tiny, it can have a crucial effect. An advantageous characteristic is handed on more effectively than a disadvantageous one, so its frequency increases in the species as a whole.

✱ If a characteristic turns out to be consistently useful over thousands of generations, its incidence will steadily expand. Instead of being an infrequent, minority feature, it becomes the norm. Thus, something that starts out as an unusual trait can eventually become an adaptation shared by a species as a whole.

it's taken me ages to get up here

PAYING FOR SEX

Some living things multiply asexually—a form of reproduction that is quick and easy, because it involves a single parent. The downside of this way of reproducing is that it normally creates no variation: the offspring are duplicates of a single parent. In contrast, sexual reproduction—because it involves two parents—creates offspring that have new combinations of both parents' characteristics. However, even in nature, there is a price to be paid for sex: it takes much more time and energy than one-parent reproduction.

13

GREAT ADAPTATIONS

* Every part of every living thing, from a seed to a complete skeleton, is the result of physical adaptations that have built up over long periods of time. Because physical adaptations are relatively easy to study—and are often well preserved in fossils—they have played a key role in evolutionary research.

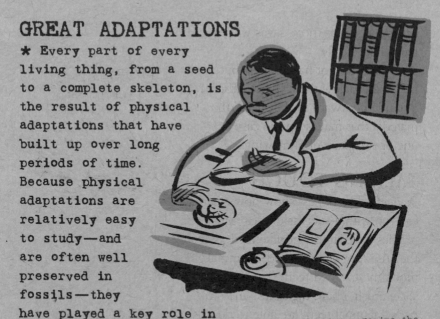

scientists examine the fossil evidence

Cones and cusps

Most reptile teeth end in a single conical point. In mammals, this plan has been modified in many ways, with several cusps on each tooth and ridges that connect them. In mammal carnivores, some of the rear teeth have become adapted to form carnassials—teeth with a scissorlike action that can slice through meat. In grazing mammals, the rear teeth have complex ridges that grind up food.

SUPPORTING ROLES

* For most organisms, physical support is essential for life. During the course of evolution, living things have come up with their own solutions to this problem, such as woody stems, hard body cases, and bony internal skeletons. Internal skeletons are all made from the same building material, but they vary from one group of animals to another. Mammal bones, for example, usually have a central cavity that is filled with marrow, but in birds the cavity is filled with air and is reinforced with fine struts to give the bone extra strength. This adaptation, which is called PNEUMATIZATION, saves weight and helps birds to fly.

* Insects do not have bones, but their body cases allow them to get airborne. Extended outward in thin flaps, the material that makes up the case produces one or two pairs of filmy wings.

TALKING TOUGH

* Teeth, like bones, show how evolution can run with a successful idea. Most reptiles have simple peglike teeth, which they use for grabbing food. Mammals have much more varied teeth that can pierce, slice, crush, chew, dig holes, and strip bark. Mammal teeth are able to do this because, unlike a reptile's, they OCCLUDE— that is, they work against each other when the mammal closes its jaws.

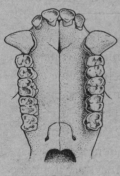

gorilla's teeth

* When an animal dies, its bones are often destroyed, but its teeth—which are harder than bone—often remain intact. *The study of fossil teeth has provided most of our knowledge about early mammals and the relationships between them.*

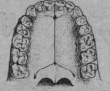

man's teeth

ANIMALS UNDER PRESSURE

In water, many soft-bodied animals stay in shape by using pressurized body fluids. The fluids press out against the animal's body wall, like air in a tire, keeping it inflated. There are also some small land animals— such as slugs and earthworms—that have HYDROSTATIC SKELETONS, though no large ones do. This is because the pressure needed to keep a large land animal "inflated" would be so great that it would explode.

ADAPTING INSIDE

* In times of drought, microscopic water animals called TARDIGRADES enter a state of deep dormancy. They can stay like this for years on end, but when water returns they rapidly come back to life. This remarkable ability, called CRYPTOBIOSIS, is an example of a physiological adaptation—one that affects the internal processes of living things.

COPING WITH CHANGE

* Unlike physical adaptations, physiological adaptations work by adjusting an organism's METABOLISM, so that its internal processes are tuned to the demands of its environment. Among other things, these adaptations help living things to cope with predictable changes—such as the passing of the seasons—and also with environmental extremes.

* Hibernation is a form of physiological adaptation. It allows a warm-blooded animal to survive winter cold by giving up the search for food

CHEMICAL MESSENGERS

In animals, many physiological processes are controlled by HORMONES—chemical messengers that are produced by glands and released into the blood. Each one acts on specific target tissues, altering the rate of different chemical reactions. The first hormone to be discovered—adrenalin— is responsible for the "fight or flight" reaction that prepares animals (and humans) to deal with danger.

I mustn't puncture this hot water bottle

some animals hibernate

16

and letting its body temperature fall close to that of its surroundings. A hibernating animal is completely inactive, and it survives on a trickle of energy released by breaking down body fat. When spring arrives, the animal's metabolism returns to normal, its temperature rises, and it groggily picks up where it left off the previous year.

llama

HIGHS AND LOWS

* Physiological adaptations help living things to survive in some of the most inhospitable places on Earth. Llamas, for example, have an unusual form of hemoglobin, the red pigment that carries oxygen in the blood. A llama's hemoglobin binds oxygen molecules even in thin mountain air higher than 13,000 feet— a height that would leave many other animals gasping for breath.

* At the other end of the spectrum, elephant seals have been known to dive to depths of more than 5,000 feet, staying underwater for over an hour. They manage this by storing oxygen in their blood and muscles, and by cutting off the blood supply to all but the most essential organs. During a dive their heart rate drops by about 90 percent, another adaptation that helps them eke out their oxygen supply while they search the ocean depths for food.

KEY WORDS

CRYPTOBIOSIS:
a state of suspended animation—the literal meaning is "hidden life"—in which an organism's internal processes drop to a fraction of their usual level

METABOLISM:
the complete range of biochemical processes that take place inside living things

ding dong!

ding dong!

Clocking off

Tardigrades, or "water bears," are probably the most resilient animals on Earth. When they dry out, their metabolic rate drops to near zero; they can survive temperatures ranging from 300° to 500°F. They can also withstand bombardment by X-rays and exposure to normally lethal poisons.

love birds

CHANGING BEHAVIOR

* Behavior is one of the least tangible features of living things. Even so, it is still shaped by evolution. When a snake strikes its prey, or a fly cleans its wings, or a male bird sets about courting a mate, all are carrying out actions that have been refined by the struggle for survival.

A HARD-WIRED WORLD

* Throughout the animal world, many forms of behavior are initiated and governed by instinct. A male weaver bird, for example, doesn't have to learn how to make its nest, even though it involves some tricky beakwork and an eye for exactly the right building materials. Instead, the necessary patterns of behavior are built into its nervous system—a biological equivalent of "hard wiring."

* The advantage of <u>INSTINCTIVE BEHAVIOR</u> is that it is prepacked and ready for use. It allows animals to respond to different situations in the most appropriate way, even though they may never have encountered them before. The downside

AIN'T MISBEHAVIN'

Plants do not behave in the same complex ways as animals, but they do respond to their surroundings. These responses usually involve specific forms of growth called <u>TROPISMS</u>—which, for example, enable plants to point their leaves toward the sun, or wind tendrils around things growing nearby. Some of the earliest research into tropisms was carried out by Charles Darwin and his son **Francis**, who described their experiments in *The Power of Movement in Plants*, published in 1881.

of instinct is that it is inflexible and can sometimes be triggered in the wrong context. A classic example of this can be seen when cuckoos trick small songbirds into raising their young. The "foster parents" instinctively respond to the young cuckoo's gaping beak by filling it with food, and they keep on responding even when the cuckoo is much bigger than they are.

the foster parents continue to feed the young monster

GETTING THE IDEA

✱ *In vertebrates, instinctive behavior is often modified by learning.* Although songbirds continue to be fooled by cuckoos, for birds and mammals, learning plays a key part in survival. The capacity to learn is itself a behavioral adaptation—one that helps animals to benefit from their individual experiences and make the most of the opportunities they encounter.

✱ *The capacity to learn is inherited, but behavior that is learned is not.* This is why all young fox cubs have to learn how to hunt, even though countless generations of fox cubs have undergone the same process, often making exactly the same mistakes.

a vine will always grow toward the sun

KEY WORDS

BEHAVIOR:
the pattern of responses shown by an animal during the course of its life; intraspecific behavior is directed to other members of the same species, while interspecific behavior is directed toward members of different species, for example predators

INSTINCT:
a form of behavior that is inherited, and which is often triggered by specific factors in the environment

EVOLUTION OBSERVED

✱ If you look at a clock, you can't see the hour hand move, even though its position is changing all the time. The same is normally true of evolution. But just occasionally —usually when humans are involved—evolution happens quickly enough to be noticed.

THE CASE OF THE PEPPERED MOTH

✱ The most famous case of "REAL-TIME" EVOLUTION began in Britain in the mid-1800s, when naturalists noticed something unusual about the peppered moth. This moth is normally pale and mottled, but from the 1850s onward, dark or MELANIC forms started to become common, particularly near cities.

✱ In the 1930s the entomologist **E.B. Ford** suggested that this change was linked to industrialization. Pale mottled moths are well camouflaged against lichen-covered tree trunks. However, Ford argued, on smoke-blackened tree trunks, dark coloring is a useful adaptation. Ford's theory was tested in the 1950s:

you can't stop me!

some bacteria are resistant to antibiotics

20

both types of peppered moth were released in polluted and unpolluted woods, and the results confirmed Ford's conjecture.

✱ In 1956, legislation was introduced in Britain to reduce industrial smoke. Research has shown that since then the balance has tipped in the other direction, and the melanic form of the peppered moth has declined steeply.

MRSA is very dangerous if it infects surgical wounds

HIGH-SPEED EVOLUTION

✱ *Because evolution involves inherited features, the shorter an organism's generation time, the faster it can evolve.* The peppered moth usually has one or two generations a year, but the fastest reproducers of all—bacteria—can cram more than 70 generations into a single day.

✱ When bacteria are cultured in a laboratory and then treated with an antibiotic, many of them will die. But, if any of the bacteria happen to be resistant to the antibiotic, this feature becomes an immense asset during reproduction. The resistant bacteria are able to multiply unhindered, and within a few days become the dominant type.

✱ In both of these cases, a change in the environment brings about a shift in the overall balance of inherited characteristics. This is what evolution is all about.

SHOT IN THE FOOT

High-speed evolution in bacteria is a growing problem in the world of medicine. After several decades of routine (mis)use, antibiotics have created strains of SUPERBUGS that are difficult to kill with drugs. One of the most problematic of these superbugs is methicillin-resistant *Staphylococcus aureus*, or MRSA. This bacterium normally lives on skin and in the lining of the nose, but it can be very dangerous if it infects surgical wounds.

GREAT TRANSFORMATIONS

* How do you convert part of an animal's anatomy used for eating food into something that picks up sound? In the case of the mammalian ear, the answer is extremely slowly and by a succession of infinitesimally small steps. This well-documented transformation of a set of body parts shows evolution at its most opportunistic, making use of whatever materials happen to be at hand.

an amazing change

mammalian earbones used to be jawbones

Movement in miniature

The three bones or OSSICLES in the inner ear are the smallest bones in the human body. Even in adults, the largest of them is not much bigger than a grain of rice. Their minute size means that they can be moved by the faintest sounds striking the eardrum.

reptile

NEW WORK FOR OLD BONES

* All mammals, including ourselves, hear sounds with the help of three very small bones located in the middle part of the ear. These three bones—called the MALLEUS, INCUS, and STAPES (or hammer, anvil, and stirrup)—conduct vibrations from the eardrum to the inner ear, where they are sensed by nerves. They act like a set of miniature levers, pivoting together in a way that makes mammalian ears highly sensitive. This arrangement of ear bones is seen only in mammals, but it did not evolve intact. Instead, it developed by the gradual "theft" of bones that once served quite different functions.

✱ In reptiles, each ear has just one bone—the equivalent of the stapes. However, unlike mammals, reptiles have several bones in their jaws. *During the gradual evolution of mammals from reptiles, some of the bones in the jaw hinge, close to the ear, began to develop a useful secondary function—that of conducting sound.* Eventually the sound-conducting bones lost their hinging role altogether and became integral parts of the ears.

CHEMICAL WARFARE

✱ *Evolutionary opportunism continues right down to the molecular level, turning substances already present in living things to new and quite different uses.* Plants, for example, form substances called SECONDARY PRODUCTS. Originally some may have been waste, while others may have played some minor role in their metabolism. However, in some cases these substances have become effective chemical weapons against plant-eating insects.

✱ This system of CHEMICAL DEFENSE has reached a remarkable degree of sophistication. Some plants can mimic insect hormones, arresting their development. As a result, the insects cannot breed normally, which reduces the chances of future attack.

PETER PAN INSECTS

The production of insect hormones by plants was first discovered by accident, when insects kept in laboratory dishes failed to develop into adults. Researchers discovered that filter paper in the dishes contained traces of an insect hormone produced by plants. Known as juvenile hormone, this substance normally helps to regulate the timing of insect development. By producing large amounts of the hormone, plants are able to prevent their insect enemies from growing up.

Venus fly trap

23

DESTINATION ANYWHERE

* Evolution is often linked with the idea of progress, and development toward particular goals. But these apparent tendencies are illusory. Instead of traveling toward a fixed goal, each species follows an evolutionary journey that has no set duration and no certain destination.

bigger and better as I get older

BIGGER IS BETTER— OR IS IT?

* *One of the best-known examples of apparent progress in evolution is provided by the horse.* Today's horses are large animals with a single toe on each foot—an adaptation that helps them to run at high speeds. Fossils of extinct horses show that today's horse has come about through a slow increase in size—from an animal about the size of a dog—combined with a gradual loss of toes.

* Looking at these fossils, it is tempting to think that evolution has worked purposefully toward a particular goal. However, this linear view of horse evolution is only part of the story. The original dog-sized horse—eohippus, or "dawn

all my nutrients are draining away

some crabs are attacked by parasites

parasitic worm

horse"—gave rise to a great variety of descendants, but only a minority are direct ancestors of the modern horse. The remainder pursued different evolutionary paths, often flourishing longer than the modern horse has, despite having small bodies and many more toes.

SHRINK TO FIT

✴ *Sometimes it can be a positive advantage to be smaller or simpler than your ancestors.* This is particularly true of parasitic animals that live permanently inside their hosts, because in this kind of sheltered environment there is no point in having complex sense organs or elaborate weapons. Instead, internal parasites concentrate on the two most essential tasks in life—feeding and reproduction.

✴ As a result, simplification sometimes goes to bizarre extremes. For example, parasites called rhizocephalans attack crabs, forming rootlike growths that spread through their bodies, draining them of nutrients. Like their victims, the parasites are crustaceans—but they look nothing like crabs; the adults having neither heads, eyes, or limbs. *In their case, evolution jettisoned a host of complex body parts that took millions of years to develop—a striking turnaround of the "onward and upward" path.*

THE PARASITE PROBLEM

In the 19th century, increasing knowledge about parasites and their lifestyles posed severe problems for naturalists who believed in a benign creator. Predators often inflict quick and relatively painless death, but parasites clearly do not fall into this category and often inflict prolonged suffering on their hosts. *For Darwin, the existence of parasites was evidence that nature was not steered by a benevolent God.*

I'm sure this fit this morning

THE END OF THE LINE

* For every species on Earth, life is destined to end in the ultimate form of failure: extinction. Some fade gradually into solitary oblivion, while others meet their end collectively, wiped out by sudden environmental change. But although EXTINCTION represents failure for individual species, it is part of the process that allows life to adapt to change.

so long, dinosaurs

OVER THE EDGE

* *Since life began, about 99 percent of the species that have evolved on Earth have died out.* Because living things do not exist in isolation, each extinction affects remaining forms of life. *The loss of a species often creates new opportunities for its competitors, while the extinction of a large number of species can radically alter the course of evolution.*

* In theory, the minimum number of individuals needed to keep most species alive is two, one male and one female. However, in practice, many species need much larger numbers in order to survive. The North American passenger pigeon, for example, was once the world's most abundant bird. In the 19th century, large-scale hunting began—but even when it slackened, the pigeon's decline continued, and in 1914 the species became extinct.

WHAT CAUSES MASS EXTINCTIONS?

Geological evidence unearthed in the 1970s strongly suggests that the dinosaurs became extinct after a meteorite struck the Earth. The causes of the other mass extinctions are less certain: cataclysmic volcanic eruptions and abrupt changes in sea level are among two of the most likely possibilities.

* Many ornithologists believe that the passenger pigeon was forced into extinction partly by its own behavior. It nested in huge colonies, and its reproductive rate seems to have fallen as those colonies shrank. Many other species depend on a certain population size, and once their population falls below a certain threshold, extinction becomes inevitable.

I'm one in 100 million

Birds by the billion

In 1810 the American ornithologist Alexander Wilson estimated the size of a flock of passenger pigeons in Kentucky at 2.2 billion birds. The last mass nesting, in 1878, covered more than 800 square miles and involved about 100 million birds. The species' subsequent extinction represents the steepest decline in recorded history.

the reign of the dinosaurs

ALL GOING TOGETHER

* During the last 500 million years, five MASS EXTINCTIONS have wiped out a large percentage of the world's living things. One of these events, 65 million years ago, abruptly cut short the long and highly successful reign of the dinosaurs. An even greater mass extinction, about 245 million years ago, killed 96 percent of marine animal species.

* *Once one of these events has occurred, life's stage is emptied of most of its actors—whereupon the remaining cast rapidly evolve into a variety of new forms.*

KEY WORDS

EXTINCTION: the disappearance of a species or group of species; once a species has become extinct, it cannot be recreated by future evolutionary processes

27

this is a mammoth task

baby mammoth

ECHOES FROM THE PAST

✱ In 1977, gold-miners in Siberia unearthed the frozen corpse of a baby mammoth. Despite being buried for more than 40,000 years, its trunk was still intact, and its flattened body was still covered by a sparse coat of red hair. Discoveries like this are rare. Instead of being frozen, buried remains are much more likely to be turned to stone.

KEY WORDS

FOSSILS:
naturally preserved remains of living things that have at some point been buried in the ground; in its narrowest sense, a fossil is something that has been mineralized or "turned to stone;" more broadly, the term is also used for non-mineralized remains

AN UNLIKELY IMMORTALITY

✱ _FOSSILIZATION_ _is a chancy business, and it works only if conditions are just right._ First, the dead remains have to escape decay. This rarely happens on dry land, but can occur in water. Fine particles of sediment can smother dead plants and animals, keeping out oxygen and preventing bacteria from breaking them down.

✱ In most cases, this kind of preservation is short-lived, and the remains are exposed, allowing decay to start. But, if they manage to avoid this fate, the stage is set for long-term chemical changes. As layers of sediment build up and become compacted into rock, the remains may become infiltrated by mineral salts. If this continues for long enough, a fossil forms.

✷ A large proportion of fossils are destroyed by rock movements and high temperatures, but some are eventually uplifted and revealed by erosion. A tiny fraction of these are discovered.

fossilization only works if the conditions are just right

THE PATCHY PAST

✷ *Fossils play a key role in the understanding of evolution, but as a history of life, the record they provide is far from complete.*

✷ One of the reasons for this is that many remains are destroyed before fossils have a chance to form. Seabed snails stand a relatively good chance of being fossilized after they die; a tree-living primate, on the other hand, does not. Its body is much more likely to be torn apart by scavenging animals when it falls to the ground.

my ambition is to be a fossil

AN OILY GRAVE

In addition to turning into stone or becoming deep frozen, methods of natural preservation include drying out (mummification), becoming entombed in amber, and (perhaps least appealing of all) being drowned and then embalmed in liquid tar. About 12,000 years ago, thousands of large mammals—including saber-toothed cats—met their end in this way at Rancho La Brea, Los Angeles, where natural tar seeps out of the ground. The animals may have been lured onto the tar by water, which forms pools on the tar's surface.

testing theories

THEORIES AND FACTS

✱ In science, theories are constantly tested. If a theory fails to match observed facts, it may have to be modified or abandoned. However if, despite rigorous testing, it continues to fit the facts, that strengthens its claim to be regarded as true. This is the current position of the theory of EVOLUTION BY NATURAL SELECTION.

KEY WORDS

HYPOTHESIS:
a proposition that can be proved or disproved by testing against observed facts

THEORY:
a speculative explanation of the underlying principles that govern any process

THE BROAD-BRUSH TREATMENT

✱ In 1905 the French mathematician and physicist **Jules Henri Poincaré** wrote that *science is built up of facts, as a house is built of stones; but an accumulation of facts is no more a science than a heap of stones is a house.* Poincaré's point was that science has to explain the facts that it accumulates and find out the principles that govern them. In mathematics and physics, these principles can be determined in an absolute way. In

EASY

it's easier
to prove a
theory in
math than
in biology

$2 + 2 = 4$

the much "fuzzier" field of
biology, things are not so
simple, because hypotheses
are not always easy to test.

✱ However, the concept of evolution
now forms an integral part of every field of
biology. It helps to explain phenomena as
diverse as the geographical distribution of
plants and animals, the development of
vertebrate immune systems, the existence
of organs that no longer seem to have any
use, and even the elaborate courtship
dances of male birds. *The process of
evolution can be used to account for events
that are taking place today, or ones that
occurred more than 3.8 billion years ago,
when life began.*

HEADLINE NEWS

✱ Today, more than 140 years after Darwin
published *The Origin of Species*, evolution
continues to make news. It has always had
its critics, particularly among people
motivated by religious beliefs, but even in
scientific circles it is the subject of
continuing debate. However, that debate is
not about whether evolution has actually
happened, but about how it works.

JUST ANOTHER THEORY

In science, some
statements are
considered to be
axiomatic, or self-
evidently true.
Examples include "two
plus two equals four"
and "air is lighter than
water." But evolution is
not self-evident, being
deduced from a large
body of observations.
Consequently, it is
regarded as a
theoretical
phenomenon, open
to disproof. Anti-
evolutionary arguments
can rarely be tested in
this way. Instead,
they are often based
on prior assumptions
that are not open
to disproof.

go on then,
prove it...

CHAPTER 1

THE EVOLUTION OF EVOLUTION

at the top at last!

* Evolution, as an idea, dates back more than 2,000 years. The Greek philosopher Empedocles, who died in about 430 B.C., suggested that the universe was in a state of gradual development, affecting living things as well as inanimate matter. His ideas were rooted in abstract thought; when other philosophers looked at the natural world in detail, many drew quite different conclusions.

THE LADDER OF NATURE

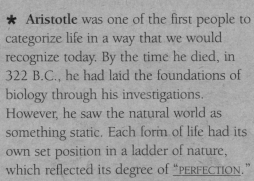

* **Aristotle** was one of the first people to categorize life in a way that we would recognize today. By the time he died, in 322 B.C., he had laid the foundations of biology through his investigations. However, he saw the natural world as something static. Each form of life had its own set position in a ladder of nature, which reflected its degree of "PERFECTION."

* At the bottom of Aristotle's ladder of nature was inanimate matter. Next came lowly or "imperfect" forms of life, such as non-flowering plants and jellyfish. Above them were more elevated organisms, such as fish and whales; and higher still, birds and mammals. The final rung of the ladder was occupied by human beings.

ARISTOTLE'S INFLUENCE

* *Aristotle's ladder had a long-lasting influence on Western thought.* It had the twin advantages of simplicity and of indeterminate length; no matter how many forms of life were discovered, the ladder could expand to fit them all. However, to modern eyes, the ladder had one strange feature: everything on it was firmly fixed in place. There was no question of anything swapping its rung for one higher up.

* By the time of the Renaissance, more than 1,800 years after Aristotle's death, his scale of nature had become absorbed into the idea of THE GREAT CHAIN OF BEING—an almost infinite gradation of species that stretched from the simplest forms of life to humanity itself.

Thales

Early thoughts on evolution

Some of the earliest Greek philosophers came up with prescient ideas about how life might have developed. **Thales** (who lived in the 6th century B.C.) and **Anaximander** (who died in 547 B.C.) both believed that living things originally arose in or from water—a conviction that corresponds with the thoughts and theories of biologists today.

I'm not very high on the ladder...

A PASSION FOR ORDER

God disposes

you are pink...
you are white...

✱ Until the 19th century, Christian ideas of creation framed European views about the natural world. The Bible describes each species as being created "after its own kind," which naturalists and theologians understood to mean fixed by divine law. In an age of increased interest in the natural world, classifying the great diversity of life became a way of testifying to the boundless powers of the Creator.

Floral sex

Linnaeus managed to scandalize some of his contemporaries by his emphasis on floral sex. He described the petals of flowers as a sweetly scented "bridal bed," where the bride and bridegroom could "celebrate their nuptials."

SIGNIFICANT SWEDE

✱ The most important figure in taxonomy—the science of classifying living things—was the Swedish botanist **Carl von Linné**, or **Linnaeus** (1707–78). An energetic traveler, Linnaeus undertook journeys throughout Europe, collecting plants in places as far apart as

Carl Linnaeus

Lapland and the shores of the Mediterranean. He was also sent pressed plants and stuffed animals by travelers overseas, which helped to give him an unprecedented overview of life as it was currently known.

✻ Linnaeus's most enduring achievement lay in drawing up an immense catalog of living things and devising a new and concise way of naming individual species. The catalog, called *Systema Naturae*, was an attempt to reflect the divine plan of creation. Initially a slim pamphlet, after many years of work it mushroomed into several volumes. ***The naming system featured in it now forms the basis of scientific nomenclature.***

more flowers please

SHIFTING GROUND

✻ To Linnaeus, the question of why some species were almost identical, save for tiny details of their structure, was not one that needed to be answered. They were like that because the Creator, in his wisdom, had made them so. But as the 18th century progressed, and yet more species were discovered, the difficulties of this view began to mount. Intrigued by the complexity of life, and the similarities and differences between species, other naturalists were feeling their way toward very different conclusions.

MAKE IT SNAPPY

When Linnaeus began his botanical career, the formal names of plants were long and cumbersome. Written in Latin, the language of European scholarship, they often amounted to a complete description of a plant and its flowers. Catnip, for example, was known as *Nepeta floribus interrupte spicatus penduculatis*. In his notes, Linnaeus reduced names like this to just two words—catnip became *Nepeta cataria* ("nepeta that attracts cats"). This personal shorthand soon caught on, and it formed the basis of *the binomial system* still used for classifying all forms of life.

"LE GRAND BUFFON"

* While Linnaeus was busy classifying the fruits of creation, a very different figure was at work in France. Rich, able, and aristocratically eccentric, Georges-Louis Leclerc, Comte de Buffon (1707-88), was one of the most remarkable gentleman-scientists of his time. His monumental illustrated encyclopedia *Histoire Naturelle*—which eventually ran to 44 volumes—contained some of the first modern speculation about evolution.

Man of many parts

Buffon's mammoth *Histoire Naturelle* was one of the most ambitious publishing enterprises of its time. The first three volumes appeared in 1749; by the time Buffon died, in 1788, a further 32 volumes had appeared. The last nine were published after his death. Buffon's other interests included mathematics, physics, and astronomy. He also dabbled in high finance.

THE PRIMACY OF SPECIES

* Buffon began his great publishing venture in 1739. *Unlike Linnaeus, Buffon rejected the idea of trying to reflect the divine plan of creation.* His views on species were perceptive but ambiguous. He recognized, for example, that species are the only biological units that have any real existence in nature; but on the subject of whether or not species were "mutable," his views were more difficult to pin down.

DEGENERATION

✱ *Buffon believed that variation could occur within a species, and he recognized that some species had vestigial features that no longer served any useful purpose.* In the fourth volume of *Histoire Naturelle*, he openly speculated about whether very similar species might have sprung from the same original ancestor. In the case of horses and asses, he wrote that their differences might be attributed to the long-standing influence of climate and of food, and to **"the chance succession of many generations of small wild horses half-degenerate, which little by little had degenerated still more... and had finally produced for our contemplation a new and constant species."** But having raised the possibility of evolution in this way, he then rejected it—perhaps, it has been suggested, to avoid controversy.

George-Louis Leclerc, Comte de Buffon

DEFINING MOMENT

Buffon was the first person to formulate the modern concept of species as units that are <u>REPRODUCTIVELY ISOLATED</u> (see page 8).

"We should regard two animals as belonging to the same species," he wrote, *"if, by means of copulation, they can perpetuate themselves and preserve the likeness of the same species; and we should regard them as belonging to different species if they are incapable of producing progeny by the same means."*

hiya cousin

some animals might have come from the same origin

NATURAL THEOLOGY

* In the same decade that saw the birth of Charles Darwin, a respected English clergyman completed a book that found lasting fame. Called *Natural Theology*, it argued that living structures were so complex and so well fitted to their tasks that they could only have been fashioned by infallible design.

complex structures must have been deliberately designed

SEARCHING FOR THE MAKER

it's about time...

Paley finds a watch

* Written by **William Paley**, an articulate defender of the Christian faith, *Natural Theology* used the complexity of nature as evidence for the existence of God and for divine involvement in shaping living things. Paley began his book with a simple example. He imagined himself, during the course of a walk, discovering a watch lying among the stones. Unlike the stones, the watch contained a collection of complex parts that worked together for a purpose. From this evidence alone, it was clear that it had been deliberately designed. Paley then extended his argument to living things: being even more complex than watches, animals and plants must also have had a maker.

I'm sure this is right...

William Paley (1743–1805) was a tutor at Christ's College, Cambridge, the same college that Charles Darwin attended in the 1820s. He was later a parish priest, a position that allowed him plenty of time for writing, before finally becoming archdeacon of Carlisle Cathedral. In addition to *Natural Theology* (1802)—which was subtitled "Evidences of the Existence and Attributes of the Deity"—his published works included *View of the Evidences of Christianity* (1794). This used the same style of reasoned argument to justify Christian beliefs, explaining that the miracles described in the New Testament were direct proof of the existence of God.

✱ Paley's logic was at its most persuasive when dealing with complicated organs such as the eye. *"Were there no example except that of the eye," Paley wrote, "it would be alone sufficient to support the conclusion which we draw from it, as to the necessity of an intelligent Creator."*

BEST BY DESIGN

✱ The idea that living things were deliberately designed found wide acceptance, both in the emerging scientific establishment and in society as a whole. Divine design was extended to include what we would today recognize as ecological relationships, and moral lessons were drawn from the behavior of animals.

✱ Because natural design was divine in origin, it was by definition perfect. For the readers of *Natural Theology*, change affecting God's creations was not only unnecessary, but unthinkable.

William Paley

39

THE FOSSIL ENIGMA

* Fossils have intrigued and mystified people for centuries. In medieval times they were thought to be the works of the devil, remains of dragons, or even imitations of living things created by some subterranean force. Their true nature, when it eventually became known, raised severe problems for the notion of a changeless natural world.

here be dragons

Jefferson on extinction

Like many of has contemporaries, Thomas Jefferson found it inconceivable that any species could have been created and then allowed to die out. "Such is the economy of nature," he wrote, "that no instance can be produced of her having permitted any one race of her animals to become extinct."

Thomas Jefferson

PUZZLES IN ROCK

* The idea that fossils have an organic origin was suggested by the Danish geologist **Nicolaus Steno (1638–86)**. However, more than a century passed before the scientific study of fossils— PALEONTOLOGY—was born. Fossil-hunting became a craze, and a number of finds were made. These included the remains of a giant sloth, found by **Thomas Jefferson**, who later became President, and the skeleton of an ICHTHYOSAURUS, discovered by the fossil collector **Mary Anning**, then just 12 years old.

✱ These fossils took some explaining. Jefferson thought that his sloth was some kind of giant lion that was rare enough to have escaped discovery. *But by the time Mary Anning discovered the first ichthyosaurus, in 1811, the "rarity" argument was in serious trouble. Even the most resolute believers in divine creation were forced to concede that, in the past at least, some forms of life had entirely died out.*

ichthyosaurus

A SLATE WIPED CLEAN

✱ *This conclusion seemed to be at odds with the accepted biblical account of the creation. Eventually, an ingenious compromise was reached. Fossils were the remains of species that had lived during earlier creations, and which had subsequently been wiped out. The species currently in existence were those formed by the most recent creation—the one that the Bible describes.*

pterodactyl

INTERPRETING BONES

The study of fossils raised a novel problem in the study of nature: **how do you identify and classify something that no longer exists?** The first pterodactyl, which was discovered in the 1780s, caused great confusion, because its bones—particularly the toothed beak and pencil-thin limbs—did not correspond to anything then known. Its true identity was established in the early 1800s by the great French anatomist **Georges Cuvier** (1769–1832). By comparing it with living species and ignoring fanciful ideas about mythical animals, he decided that it had to be a flying reptile. Cuvier also helped to clear up the mystery surrounding Mary Anning's ichthyosaur—deciding that it, too, was a reptile and not a fish.

RECORD IN THE ROCKS

* By the beginning of the 19th century, two leading geologists, the German <u>Abraham Werner</u> (1749-1817) and Englishman <u>William Smith</u> (1769-1839), had helped to establish that rocks are laid down as distinct layers or strata, which follow a set sequence in time. From this, it became clear that fossils, too, are arranged in a definite order, forming a record of life that stretches back into the past.

William Smith

LAYERS OF LIFE

the answer's in the rocks

* William Smith worked as a surveyor in the construction of mines and canals. As a result, he was well placed to investigate geological strata and the fossils that each one held. *In his principle of superposition, he asserted that—as long as they have not been disturbed—higher strata must be more recent than ones that lie beneath them. He also showed that sedimentary rocks could be identified by the fossils that they contain.*

* *By firmly linking fossils with past geological eras, Smith helped to transform them from random relics into*

Noah's ark

pages from the history of life. But it was a history with peculiar features: between neighboring strata, rocks and their fossils often abruptly changed, for no obvious reason. How had these changes occurred?

A SERIES OF CATASTROPHES

✱ To the members of the CATASTROPHIST school of thought, among them Abraham Werner, these sudden changes were caused by immense cataclysms—such as the biblical flood. *Catastrophism fit in well with the idea of successive creations, each with its own set of species, and of the slate being wiped clean when each one came to an end.* But, as knowledge of strata increased, the number of catastrophes required grew and grew. To some geologists, catastrophes started to look less like exceptional events and more like commonplace features of the Earth's geological history.

Uniformitarianism

The idea that "catastrophes" were quite common events formed the core of the "uniformitarian" view. The principle of uniformitarianism—that the present is the key to the past—did away with the need for cataclysmic change. Instead, it emphasized the importance of everyday processes that continued for immense spans of time. It was to have a profound influence on geology—and on the thought of Charles Darwin as he pondered the history of life.

fossils began to give up their secrets

HOW OLD IS THE EARTH?

* According to uniformitarianism, the surface of the Earth is shaped by everyday geological processes. Because these processes work very slowly, it follows that the Earth must be extremely old. Today this is accepted as a scientific fact, but back in the 18th and 19th centuries it was a controversial claim that had major implications for the study of prehistoric life.

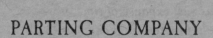

when did it all begin?

PARTING COMPANY

only a few thousand years old!

* The dispute about the Earth's age stemmed from a growing divergence between the biblical record of creation and newly developed scientific ideas. *In the 1700s several new theories were put forward about how the Earth might have formed, all of them at odds with the accepted idea that the Earth was just a few thousand years old.* In the first volume of his *Histoire Naturelle*, published in 1749, **Georges-Louis de Buffon** (see page 36) suggested that the Earth might have been created from a ball of white-hot matter that had steadily cooled over a period of about 70,000 years. Off the record, he conceded that even this figure was probably an underestimate.

✱ *To minimize the chances of dispute with the Church, attempts were made to reconcile these ideas with the biblical creation story. One strategy, which Buffon himself used, was to assume that the biblical "days" of creation were not days in the literal sense. He divided the Earth's 70,000-year history into six epochs, equivalent to the six days of creation.*

TIME AND LIFE

✱ *By the 1830s, most geologists accepted that the planet had been in existence for tens or perhaps hundreds of millions of years, the precise age being uncertain. Given this vast expanse of time, it was possible that incremental changes had modified living things, just as different incremental changes had modified the surface of the Earth.*

✱ This realization removed an important stumbling block in the way of evolution. However, in later years, controversy about the Earth's age was reignited, and it escalated into the source of what Darwin called one of his "sorest troubles."

The biblical age of the earth

In the 1650s, after many years studying the genealogies recorded in the Old Testament, James Ussher, Archbishop of Armagh, pronounced that the world was created in the year 4004 B.C. His contemporary John Lightfoot, Vice-Chancellor of Cambridge University, was even more precise, calculating that the final act of creation—the genesis of the human race—occurred at 9 o'clock on the morning of Sunday, October 23.

archbishops and scientists argued about creation

45

diplodocus

Dinner in the dino

On New Year's Eve in 1853, a group of 21 prominent scientists dined inside a model iguanodon that was about to be installed in London's Crystal Palace. The life-sized model was designed by Richard Owen, who was among the guests.

DINOMANIA

* When the first major dinosaur finds came to light during the first half of the 19th century, the public reaction was one of amazement that such gigantic creatures could have existed, only to have disappeared. In scientific circles, "dinomania" helped to fan a growing debate on the possibility that species might be capable of change.

TELLTALE TEETH

* The first remains to be linked with dinosaurs were fossilized teeth, found in 1822 by the English doctor **Gideon Mantell (1790–1852)**. At this time dinosaurs were unknown, but Mantell correctly deduced that the teeth belonged to a lizardlike reptile, similar to an iguana. He called it an IGUANODON ("iguana tooth"), and estimated that it must have been about 40 feet long.

* It was not until 1841 that the British anatomist **Richard Owen (1804–92)** coined the collective

they keep on finding bits of me

Richard Owen
studied dinosaur
remains

name "dinosaurs" (meaning "terrible lizards") for this new group of reptiles. Owen studied dinosaur remains in detail, and announced that they were not only larger than modern reptiles but—significantly—also more advanced.

* *By the 1840s, evolution was being openly discussed, and the idea of progress was almost invariably linked with it. By showing dinosaurs to be "superior" to modern reptiles, rather than more primitive, Owen believed he had evidence that undermined evolutionary theory.*

* Today evolutionary theory holds that species can adapt by becoming either more simple or more complex, so Owen's argument no longer has the force it did. But the "tendency to progression" was an important idea in evolution, thanks largely to the work of one man—the French naturalist **Jean-Baptiste de Lamarck** (1744–1829).

Richard Owen (1804–92) is often dismissed as the "bad guy" in the story of evolution—but, while he clashed fiercely with Darwin about natural selection, his views of evolution itself seem to have been less clear-cut. A highly gifted anatomist, he recognized that many of the structures seen in living things are variations on underlying themes (see page 12). Early in his career he seems to have been firmly anti-evolutionist, but some of his later writings suggest that he gradually changed his mind. However, unlike Darwin, Owen seems to have believed in a limited kind of evolution that followed a divine plan.

skull of a bird

LAMARCK

✱ Jean-Baptiste de Lamarck (1744-1829) was the first person to set out a theory of evolution that explained how and why change occurs. He believed that living things had a natural tendency to progress, and that they could pass on useful features developed during the course of their lives.

Jean-Baptiste de Lamarck

Jean-Baptiste de Monet, Chevalier de Lamarck, came from a family of impoverished aristocrats. After attempts at various careers, he was appointed botanist to Louis XVI. He survived the turmoil of the French Revolution to become a professor of zoology in Paris, where he devised his evolutionary theory.

The role of nervous fluid

Lamarck believed that an elusive "nervous fluid" was responsible for making living structures develop when they were subjected to increased use. Parts of the body that were used most often would attract large amounts of the fluid, increasing their size. The opposite effect also held true—explaining how body parts such as wings degenerated when they fell out of use.

THE ONWARD URGE

✱ In *Philosophie Zoologique*, published in 1809, Lamarck claimed that simple forms of life constantly arise from non-living matter and that they gradually evolve and become more complex. *According to Lamarck, this process is driven by the needs of each organism as it strives to fulfill its way of life.*

✱ The giraffe provides a classic example. Long ago, the ancestors of giraffes began to eat the leaves of trees. As they stretched upward to reach branches, their necks became elongated; and this characteristic was passed on when they bred. After many generations this process produced today's long-necked giraffes.

this neck's very useful

how did giraffes get their long necks?

FLAWED THEORY

✱ At first glance, Lamarck's explanation seems quite plausible, but a moment's thought soon reveals some flaws. *If characteristics acquired during life really were passed on, we would soon notice the effects.* For example, trees bent over by strong winds would produce lopsided trees; and people with suntans would produce suntanned children. Neither of these things happens, because changes like these do not affect the biological blueprint that is coded by **genes**.

✱ Nonetheless, Lamarck's ideas proved persistent. *Until Darwin published his account of evolution by natural selection, Lamarck's theory was the only detailed argument of how change might occur across generations.*

curlew

49

VESTIGES OF CREATION

* The year 1844 saw the publication of a small and eccentric book, *Vestiges of the Natural History of Creation*. Among other things, its author—who remained anonymous at the time—proposed a mechanism for evolution. As events turned out, the writer had good reason to withhold his name. *Vestiges* was roundly denounced by scientists and churchmen alike.

Robert Chambers

In *Vestiges*, Chambers had some harsh words for the belief that God specially created each species. *How can we suppose that the august Being... was to interfere personally and on every occasion when a new shell-fish or reptile was to be ushered into existence...?* He concluded that *the idea is too ridiculous for a moment to be entertained.*

hum...now what would a shrimp look like?

HEAD ABOVE THE PARAPET

* *Vestiges* was written by **Robert Chambers**, a publisher and also a keen amateur scientist. In a chapter entitled "Origin of the Animated Tribes," Chambers asserted that the Earth was not specifically created by God, but was formed by laws that expressed the Creator's will. He then turned his attention to living things:

"*That God created animated beings... I at once take it for granted. But in the particulars of this so highly supported idea, we surely see cause for some reconsideration.*"

* Chambers' "reconsideration" led to a theory of evolution that was prompted by environmental factors. Like Lamarck, he believed that evolution was progressive, and he thought that it followed a predetermined plan. As far as animals were concerned, he believed that the end result of that plan was man. He also believed in spontaneous generation—the idea that life could arise from nonliving matter.

A HOSTILE RESPONSE

✱ In the liberal atmosphere of France, Lamarck's views on evolution (see page 48) were greeted with scepticism mixed with indifference. In Britain, however, a far more hostile response awaited anyone who dared to espouse evolution in print. Adam Sedgwick, Professor of Geology at Cambridge, wrote a damning 85-page review, claiming that the country's "glorious maidens and matrons" needed to be protected from such ideas. Members of the clergy were outraged, in particular by the suggestion that humans had arisen from animals.

✱ Despite this barrage of criticism, *Vestiges* was a successful publishing venture and ran to many editions. But it served as a warning. *For one of Chambers' contemporaries— Charles Darwin—the dangers of advocating evolution were now plain.*

Leaping into life

Until the early 1800s, it was believed that some forms of life could arise spontaneously from non-living matter. Maggots, for example, were supposed to arise from putrid meat, while soiled underwear could produce mice from grains of wheat. Chambers credited reports that electricity could "create" fully formed insects. The theory of spontaneous generation was not finally abandoned until 1860, after conclusive experiments by the French microbiologist Louis Pasteur.

this is an outrage

the clergy were shocked by Chambers' book

CHAPTER 2

I can't see
myself as a
doctor...

CHARLES DARWIN

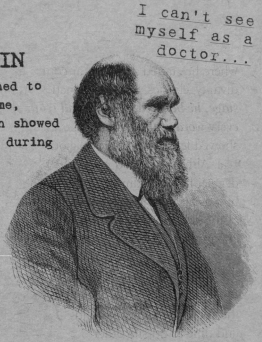

* For someone destined to
achieve worldwide fame,
Charles Robert Darwin showed
few signs of promise during
his early life. Born
in 1809 into a
prosperous and
talented family, he
initially trained
to be a doctor, but
within two years
gave up medicine and
decided to become a
clergyman instead.

Darwin the dropout

Charles Darwin's school
career was far from
distinguished. At the age
of 16—just before he was
sent off to study
medicine—his father told
him that "you care for
nothing but shooting,
dogs and rat-catching, and
you will be a disgrace to
yourself and all your
family." Paradoxically, one
of the reasons Charles
Darwin gave up medicine
was that he could not
stand the sight of blood.

THE LEISURED LIFE

* For Charles Darwin, money was not a
problem. His father was a successful
doctor, while his mother—who died when
he was just eight—was a daughter of
Josiah Wedgwood, an industrialist who
made a fortune in the pottery business.
The Darwin family's comfortable
circumstances meant that Charles never
had to earn his living, which
was of enormous value when
his interest eventually focused
on evolution.

* In 1827, after his
unsuccessful foray into
medicine, Darwin moved to
Christ's College, Cambridge,

spider

where he started a three-year course in divinity, classics, and mathematics. *At this stage, he believed in the literal truth of every word of the Bible;* but, even so, he showed few signs of any religious zeal. Although he did get his degree, he later admitted that he was far more interested in riding and shooting than in going to lectures.

...no neither can I

THE TRAVEL BUG

✱ Despite his fondness for the sporting life, Darwin did find time to pursue other interests. He attended public lectures on botany given by Professor John Henslow, and he became an avid collector of beetles and other forms of insect life. He was also an armchair traveler, avidly reading accounts by Alexander von Humboldt, the celebrated German explorer and naturalist, of his five-year journey throughout Central and South America. *Von Humboldt's descriptions of tropical wildlife fired Darwin's imagination; he craved to see the tropics for himself.*

✱ After graduating, Darwin remained at Cambridge to study geology, but he had been bitten by the travel bug. Eventually, an opportunity arose through **John Henslow's** help. *This was to be the only major journey that Darwin ever undertook, and it would shape the rest of his life.*

John Henslow

VOYAGE OF THE *BEAGLE*

✱ On December 27, 1831, the naval survey ship HMS *Beagle* set sail from Devonport in southwestern England. Its mission was to chart the coastal waters of Patagonia, Chile, and Peru, during a voyage that would eventually last five years. In addition to the ship's usual complement of officers and ratings, it carried a gentleman-naturalist: young Charles Darwin, ready for adventure.

LIFTING LAND

As the *Beagle* made its way up the west coast of South America in 1834, Darwin experienced a major earthquake at the Chilean port of Valdivia. When the quake was over, he noticed that the coast had been raised by about a yard, lifting mussels and other animals out of the water. He had already found fossil shells high up in the surrounding mountains and became convinced that they had been raised in a series of many small steps. It marked his acceptance of the uniformitarian view of geology (see page 43), an important break with the past.

TRAVELING COMPANIONS

✱ Darwin had been invited to join the *Beagle* by the ship's captain Robert Fitzroy, who wanted an educated companion to relieve the tedium of many months afloat. Darwin's father was initially against the

Peru

Chile

Patagonia

South America

the naval
survey ship
Beagle set sail
from Devonport

idea, although he eventually relented, but Darwin turned out to be a bad sailor, and was often seasick.

* *Darwin's post was an unpaid one, and he was treated more as a passenger than a member of the crew. He made good use of his freedom. While the* Beagle *carried out soundings, often tracking back and forth along the same length of coast, Darwin was able to journey ashore for days at a time, making observations about geology, plants, and animals, and collecting specimens.*

MYTH IN THE MAKING

* When he set sail on the *Beagle*, Darwin did not believe in the truth of evolution. Despite accounts of his "conversion" during the voyage, today's historians believe that he had still not changed his mind five years later, when the *Beagle* eventually returned to England. But unlike Fitzroy, Darwin was at least open to evolution as a theoretical idea.

Tropical delights

In February, 1832, the *Beagle* made a landfall in Brazil, allowing Darwin to explore a tropical forest for the first time. He was ecstatic:
"Delight... is a weak term to express the feelings of a naturalist who, for the first time, has wandered by himself in a Brazilian forest. The elegance of the grasses, the novelty of the parasitical plants, the beauty of the flowers, the glossy green of the foliage, but above all the general luxuriance of the vegetation, filled me with admiration'"

Charles Darwin, *Journal of Researches*

ISLANDS APART

* After tracking northward
along the coast of Chile and
Peru, the *Beagle* struck out
westward toward the remote
Galapagos Islands. Arriving
in September 1835, it spent
a month charting the
islands' harsh volcanic
shores. During his time in the
Galapagos, Darwin made some of
his most important discoveries
about the local wildlife—
discoveries that later helped
him to understand the way
evolution works.

the Galapagos
Islands

THE FINCH FACTOR

* Darwin was fascinated by the strange
and sometimes bizarre animals he
encountered—including giant tortoises,
marine iguanas, and flightless cormorants.
As the *Beagle* shuttled from one island to
another, he discovered some curious facts
about the islands' wildlife.
Instead of sharing the
same animals, each
island seemed to have
its own particular
"brands." *For
example, there
were over a
dozen kinds of*

Darwin

THE GALAPAGOS

Lying about 600 miles
west of South America,
the Galapagos Islands
are a cluster of extinct
or dormant volcanoes
separated by deep
ocean channels. The
climate is dry for most
of the year, and on low
ground the islands
consist of little more
than scorched basalt
lava. *Darwin realized
that the Galapagos—
which are geologically
very young—must have
been colonized by
animals and plants
arriving from outside.*
In the case of the
finches, he decided that
these colonists must
have come from the
Americas.

finches. They all had slightly different beak shapes, which enabled them to live on different foods, and they all lived in different places.

✱ These differences fascinated the young English naturalist. In his account of the *Beagle's* voyage, he wrote: *"I never dreamed that islands about 50 or 60 miles apart, and most of them in sight of each other, formed of precisely the same rocks, placed under a quite similar climate, rising to nearly equal height, would have been differently tenanted..."* It was a puzzle that set his mind working.

MISSING LUGGAGE

✱ *Darwin did not realize the significance of his findings while he was on the islands; and by the time he left, his collection of specimens was far from complete.* He did not collect any of the different giant-tortoise shells, and he was missing a number of the finches. To make matters worse, he had not noted where his collected finches came from, because at the time he did not think the information was worth recording.

✱ As the *Beagle* headed out across the Pacific, he may well have regretted this oversight, because the Galapagos finches begged many questions. They were to become a central piece of evidence in his new evolutionary theory.

Darwin's finches

With the help of the British ornithologist John Gould, Darwin later established that there were 13 species of Galapagos finches, and that they could be divided into four groups. Some had heavy beaks shaped for cracking open seeds—just like typical finches elsewhere—but others had much slenderer beaks for catching insects. One of these insect-eaters had even found a way of extracting insects from crevices in bark, using cactus spines to pry them out.

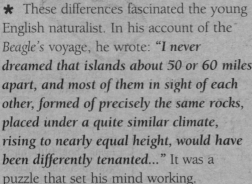

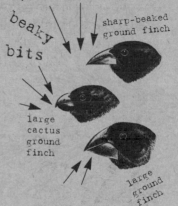

beaky bits

sharp-beaked ground finch

large cactus ground finch

large ground finch

HOMEWARD BOUND

* After leaving the Galapagos Islands, the *Beagle* crossed the Pacific and Indian Oceans, before rounding the Cape of Good Hope and zigzagging its way back to England. By the time it docked on October 2, 1836, Darwin had already had a year to ponder the discoveries he had made during the five-year voyage. Slowly and tentatively, he was feeling his way toward some potentially explosive conclusions.

life is not all that it seems!

Darwin on board the *Beagle*

DARWIN'S OTHER BESTSELLER

Darwin's account of his work on the *Beagle* was originally published as part of the expedition's *Journal of Researches*. Unlike the rest of the Journal, it had a huge appeal for Victorian armchair travelers, and it was later published separately as *The Voyage of the Beagle*.

LAVISH LIFE

* *According to the conventional views of the time, all animals and plants had been created to fit the conditions that they lived in. However, Darwin reasoned, if this was so, the same species should crop up in the same habitats all over the world. But his experiences on the* Beagle *had showed him that each part of the world seemed to have its own species.*

* For example, on the South American pampas Darwin had often seen large flightless birds called greater rheas. Their habitat was like that of the African ostrich, and their anatomy was also similar.

the Beagle

But ostriches were only found in Africa, while rheas were confined to the Americas. Why were there two species when one would do?

✱ In the remote Galapagos archipelago, nature seemed even more extravagant. Here, each island often had its own unique forms of life. *Had the Creator gone to all the trouble of making 13 separate species of finch solely for a scattering of rocky outcrops in the vastness of the Pacific?*

KEEPING MUM

✱ In 1845, with the publication of his *Journal of Researches*, Darwin let his thoughts on the finches be heard.

✱ *"Seeing this gradation and diversity of structure in one small, intimately related group of birds, one might really fancy that from an original paucity of birds in this*

neck

feathers

legs

ostrich

archipelago, one species had been taken and modified for different ends." Darwin was describing a process known as ADAPTIVE RADIATION. How it occurred, he was not yet prepared to say.

tortoise

The origins of atolls

By the time Darwin returned to England, his work was well known in scientific circles, but he was thought of more as a geologist than a biologist. During the *Beagle*'s homeward run, he had carried out surveys of coral reefs in the Indian Ocean and had become convinced that atolls form where islands are slowly subsiding into the sea. Darwin's work on reefs helped to establish him as a serious scientist.

THE LONDON YEARS

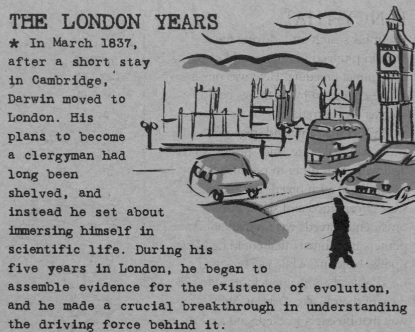

* In March 1837, after a short stay in Cambridge, Darwin moved to London. His plans to become a clergyman had long been shelved, and instead he set about immersing himself in scientific life. During his five years in London, he began to assemble evidence for the existence of evolution, and he made a crucial breakthrough in understanding the driving force behind it.

I've put on too much weight to be a racing pigeon

SECRET JOTTINGS

* Darwin was a prolific writer, and his letters and notebooks, many of which still survive today, show how he steadily moved away from the current thinking of his time. In 1837 he began the first of several notebooks on the SPECIES QUESTION. From this notebook, it is clear that he had taken a decisive step forward by accepting that species are capable of transmutation, or change, but he still could not explain exactly what made this change occur.

Married life

In January 1839 Darwin married his cousin Emma, consolidating his links with the Wedgwood family. The couple eventually had ten children, and their relationship was deep and affectionate. However, Emma was a devout Christian, and she was often troubled by her husband's work. These differences of opinion gave Darwin a hint of the difficulties that would arise when he eventually made his thinking public.

FANCY THAT

✶ In his search for a driving force, Darwin plunged into the world of plant and animal breeding. Here was one area where change in living things could be seen as it occurred.

✶ In Victorian England, pigeon breeding was a national craze, and "pigeon fanciers" vied with each other to produce the best examples of each breed. Racing pigeons were streamlined and sleek, but ornamental breeds were remarkably varied: pouters had throats that could be blown up like balloons, while jacobins had a hood of feathers that stuck up from the back of their necks. Darwin studied the way that breeders worked, and sent them detailed questionnaires. *From his research, he concluded that pigeon fanciers had created this wealth of breeds from a single ancestral species by selecting which bird bred with which.*

✶ Selection seemed to be the key to evolutionary change, but it was one thing for a plant or animal breeder to artificially select individuals that had useful characteristics, and quite another for it to happen in the wild. *How did selection occur in nature, where there was no one to control which individual bred with which? For a while, Darwin was mystified, but in 1838 he stumbled on the answer.*

Yes, but mine is much faster

the Victorians liked breeding pigeons

61

Thomas Malthus

THE STRUGGLE FOR EXISTENCE

✱ In 1798, an English academic called Thomas Malthus published his *Essay on the Principle of Population*. It argued that human numbers tend to increase geometrically, which means that they eventually outstrip their food supply. Malthus wrote that the ensuing competition for limited resources must inevitably create a struggle for existence.

Natural birth control

All living things have the potential to replace themselves many times over. Female houseflies, for example, lay hundreds of eggs. If all their offspring survived, the housefly population would increase geometrically, and soon the world would be swamped by flies. However, like other living things, houseflies have a relatively static population. In the struggle for limited resources, relatively few manage to survive.

THE LIGHT DAWNS

✱ Although Darwin had heard of Malthus' essay, he did not get around to reading it until 1838. When he finally did, its effect was electrifying. After reading it, he wrote in his autobiography that *"being well prepared to appreciate the struggle for existence... it at once struck me that under these circumstances favourable variations would tend to be preserved and unfavourable ones destroyed. The result of this would be the formation of a new species."*

✱ *Competition, rather than deliberate selection, seemed to be the force that steered change in living things.* Poor performers would be weeded out in the struggle, while the ones that fared best—by

ding

Malthus gave
Darwin the
clue he needed

The Reverend *Thomas Malthus (1766–1834)* was a clergyman, philosopher, and economist. His *Essay on the Principle of Population* was intended as a warning against the effects of welfare provision for the poor, which he believed would ultimately backfire by increasing the number of mouths that had to be fed. Over time, Malthus' name became associated with repressive measures designed to make the poor independent, but Malthus actually advocated more humane solutions to the problem of poverty, such as education to reduce family sizes.

leaving the most offspring—would pass on their characteristics most effectively. *By applying Malthus' ideas to the natural world, Darwin had found the evolutionary driving force he had been searching for:* NATURAL SELECTION.

THINNING THE RANKS

* Despite Darwin's flash of insight, the idea of natural selection was not new. Lamarck had written about it; and so had William Paley, whose work Darwin had studied at Cambridge. But to Lamarck and Paley, this kind of selection simply weeded out individuals that were not fit to survive. It left the survivors unchanged, in the form in which they were originally created.

* Darwin's ideas were very different, because he saw natural selection as a force for change. By weeding out the least fit, it constantly adjusted the characteristics of future generations.

survival of
the fittest

COUNTRY RETREAT

Downe House

* In 1842, Darwin decided that the time had come to move out of "dirty odious London." Although he was hard at work, he did not feel that he had enough evidence to publish his ideas, and the pressure of self-imposed secrecy was beginning to affect his health. The family settled in a small country house at Downe, in Kent, where Darwin continued his research in an atmosphere of peace and privacy.

THE SOURCE OF VARIATION

Throughout his work on evolution, Darwin believed that variation in living things was probably caused by natural changes in their habitats. Subsidence or flooding, for example, would have forced living things to alter their ways, creating physical changes that were then passed on. This clearly echoes the ideas of Lamarck (see page 48), showing that Darwin was not as completely at odds with Lamarckism as is often supposed.

PUTTING THE RECORD STRAIGHT

* Shortly before arriving in Downe, Darwin completed a 35-page summary of his ideas, written in pencil. Two years later, he expanded this into a 230-page essay. In this expanded account, he envisaged evolution as a process that took place only in response to changes in the environment. At this stage, he believed that these periods of change were interspersed by long intervals of stability, in which living things remained practically the same.

* Darwin's evolutionary theory was gaining credibility, but it still had a number of major loopholes. One of the most glaring concerned the problem of

variation. In the 1840s, no one had any clear understanding of why offspring often differ from their parents, or how characteristics are passed on. The problem was a big one—because without variation, natural selection has nothing on which to act. It was not solved in Darwin's lifetime, and it continued to perplex him in years to come.

say cheese

DARWIN'S MYSTERY ILLNESS

✱ Despite the congenial surroundings at Downe, Darwin's health began to decline. He was quickly exhausted by work, and he felt that his long-term prospects were poor. He even made arrangements to have his essay published posthumously if he died, fearing that his work might otherwise be lost.

✱ In the end, Darwin lived for another 40 years, and his mysterious chronic illness has been the subject of much speculation. Many historians think that it was probably some kind of psychological disorder, caused by a fear of the hostile reaction his work would eventually unleash.

Darwin suffered from a mysterious chronic illness

Night bite

One theory about Darwin's illness is that he contracted Chagas' disease, which is transmitted by bloodsucking bugs. In *The Voyage of the Beagle,* he describes being bitten: *"At night I experienced an attack (for it deserves no less a name) of the Benchuga... the great black bug of the Pampas. It is quite disgusting to feel soft wingless insects, about an inch long, crawling over one's body."* However, this theory does not account for the timing of his illness—he had repeated bouts of ill-health long before he was bitten.

Joseph Hooker

MURDER MOST FOUL

* Anxious about the potential impact of his ideas, Darwin held back from committing anything to print. However, in private, he kept in touch with many leading scientists of the day. This Victorian equivalent of networking enabled him to test reactions to his work and helped prepare the ground for the time when his theory would eventually be published.

ADMISSION OF GUILT

* In 1844—the year that his essay was completed—Darwin wrote to **Joseph Hooker**, an eminent botanist who was by now a close personal friend. *He told Hooker that he no longer believed that*

species were immutable, and also said that he had discovered *"the simple way by which species become exquisitely adapted to various ends."* But, despite his growing confidence in his ideas, Darwin felt that there was still much work to do, and that it was not yet time to publish. He wrote to Hooker that *simply believing in evolution seemed like confessing to a murder*.

SHOT ACROSS THE BOWS

✱ 1844 was also the year in which *Vestiges of the Natural History of Creation* appeared (see page 50). Darwin found much to criticize in the book, and the furor that greeted it no doubt sharpened his resolve not to release his own theory prematurely. As the years went past, Darwin became involved in work that became almost obsessively detailed—anatomical studies of pigeons, monographs on barnacles, studies of the way plants and animals manage to spread. By the mid-1850s his friends were warning him that, if he did not publish soon, there was a danger someone might get there before him.

✱ In 1856 he began work on what he referred to as his "big book," which was to be called *Natural Selection*. However, two years later—while the book was still in progress—his friends' worst fears were realized. Someone else was on the trail.

Murky waters
During the mid-1800s, little was known about the biology of barnacles. Biologists had only recently discovered that they were crustaceans—relatives of animals such as crabs and lobsters—and not mollusks, as was formerly thought. Darwin dissected thousands of barnacles from all over the world, and his barnacle work became such a feature of life at Downe that one of his sons once asked a friend where his father "did his barnacles."

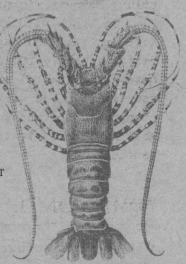

lobster

NEWS FROM THE EAST

✱ On June 18, 1858, a letter arrived at Downe House. It was from Alfred Russel Wallace, a naturalist who had

been working in Southeast Asia, and it contained a manuscript that Wallace hoped to have published. As Darwin read the paper, he realized that he had been scooped: it exactly mirrored his own thinking on natural selection.

Charles Lyell

Charles Lyell (1797–1875) was one of Victorian Britain's most distinguished geologists. He helped to establish the principle of uniformitarianism (see page 43), which influenced Darwin's thinking on evolution. However, despite his support for Darwin, Lyell remained doubtful about evolution—particularly when it included human beings.

PANIC ATTACK

✱ Wallace's paper was called *On the Tendency of Varieties to Depart Indefinitely from the Original Type. In a handful of pages, it summarized the idea of the struggle for survival in nature and*

I've discovered evolution! I should get malaria more often

explained how, in a world of variation, only the best-adapted individuals would survive. Darwin was stunned. He had met Wallace and knew that he believed in the possibility of evolution. But Wallace had gone one step further, hitting on exactly the same driving force that Darwin had proposed.

✱ Darwin sent the paper to a friend, Charles Lyell. With it, he included an agonized and heartfelt letter. *"I never saw a more striking coincidence,"* he wrote. *"All my originality, whatever it may amount to, will be smashed... ."*

THINKING ALIKE

✱ The coincidence was even deeper than it first appeared: like Darwin, Wallace had based the idea of natural selection partly on the work of Thomas Malthus. While ill with malaria on an island near New Guinea, Wallace had a moment of inspiration, linking population growth and the struggle for survival. *Whereas Darwin had been at work on the topic for years, Wallace wrote his paper in just a few days. The obvious person to send it to was Darwin himself.*

✱ After the shock subsided, Wallace's move had a catalytic effect. Darwin did not want to undermine Wallace's achievement, but neither did he wish to throw away his own more detailed research. Eventually, his friends convinced him that he had to act. *It was time to make his theory public.*

Alfred Russel Wallace (1823–1913) was a naturalist and collector of botanical and zoological specimens. He made a collecting expedition to the islands of modern-day Malaysia and Indonesia in 1854. In addition to co-discovering natural selection, Wallace was a pioneer in biogeography, the science of plant and animal distribution. He identified the clear-cut boundary that separates the species of Australasia from those of Southeast Asia. This boundary, which runs between the islands of Bali and Lombok, is known as "Wallace's Line."

Wallace's line

DARWIN GOES PUBLIC

✶ Acting on the advice of Hooker and Lyell, Darwin arranged for a joint presentation of his work and Wallace's. On July 1, 1858, contributions from the two men were read aloud during a meeting of the Linnaean Society in London. Despite their revolutionary content, the papers attracted only a modest amount of interest.

it's like this

Darwin at the Linnaean Society

ME FIRST

✶ The meeting heard Wallace's paper and an extract of Darwin's essay that had been written over a decade earlier in 1844. Sandwiched between them was a letter that Darwin had written in 1857 to

---

WHAT HAPPENED TO WALLACE?

Wallace's eclipse at the Linnaean Society meeting did not signal the end of his career. He did further work on natural selection, writing several books on the subject, but his greatest contributions to biology continued to be in the field of biogeography. However, Wallace's interests roamed in a way that Darwin's never did. He became involved in socialism and women's rights, and also in spiritualism. Like Darwin, he believed that human beings were the products of evolution, but felt that the human spirit was not produced by natural selection and had to have a supernatural origin.

the American botanist Asa Gray, discussing the question of whether natural selection inevitably tended to make species diverge. The implication was that Darwin had a prior claim to be considered the "discoverer" of evolution by natural selection. From this moment on, the whole subject of evolution was to become synonymous with Darwin alone.

✱ However, if Darwin considered his ideas explosive, the bomb did not go off immediately. In his year-end summary of 1858, the Linnaean Society's president remarked that he did not have any striking discoveries to report.

HARD GRAFT

Darwin realized that he would have to change tactics. He no longer had time to complete *Natural Selection*. What was needed was a smaller book that would sum up the evidence in a way that non-specialists could understand.

✱ Darwin threw himself into the work, and the book was finally ready for the printers in October 1859. It was called *On The Origin of Species by Means of Natural Selection, or the Preservation of Favoured Races in the Struggle for Life*. Better known simply as *The Origin of Species*, it has remained in print ever since.

KEY WORDS

DARWINISM:
word originally coined in the late 1700s, when it was used to describe the ideas of Charles' grandfather, Erasmus Darwin (see page 53); subsequently it came to be associated with Charles' theory that evolution is driven largely by natural selection

NEO-DARWINISM:
describes the modern version of Darwinist theory, updated to take account of discoveries in genetics.

Darwin's bombshell

INSIDE *THE ORIGIN*

* Compared with other books that have revolutionized scientific thinking, *The Origin of Species* is a surprisingly straightforward read. Using a minimum of scientific jargon, Darwin lays out the evidence for believing that living things evolve and explains how natural selection makes species change.

Total sell-out

The original print run for *The Origin of Species* was only 1,250 copies, priced at 15 shillings each—a sum equivalent to at least $50 today. The publisher felt that it was unlikely to generate much demand, but his pessimism was unwarranted: the entire print run was bought up by booksellers on the first day.

existence
natural
selection

WOW!

SPECIES IN WAITING

* Darwin described *The Origin* as "one long argument." To make the argument easier to follow, he decided to begin with something his readers would have found familiar—the range of variation that can be seen in domesticated plants and animals. After discussing the features of different breeds—his favorite domestic animal, the pigeon, is paid particular attention to—Darwin moves on to explain how these breeds have come about. The answer is artificial selection, a process in which humans decide which individuals should be allowed to reproduce.

✳ The second chapter of *The Origin* moves away from the human world to investigate variation in the wild. Here Darwin discusses the problems involved in defining exactly what a species is, and in separating species from mere varieties. At this point, he parts company with most Victorian biologists. *He writes that, instead of being fixed, "well-marked varieties" are species-in-waiting—part of the way along the road to becoming species in their own right.*

I'm Darwin's favorite domestic animal

SLOGGING IT OUT

✳ Chapter 3 deals with the ideas Darwin had developed after reading Malthus. He writes that **"a struggle for existence inevitably follows from the high rate at which organic beings tend to increase,"** even if that being reproduces as slowly as an elephant. He then goes on to outline factors that limit species numbers, including the effects of overcrowding, attacks by predators, extreme cold, and drought. At the outset, he stresses a point that is often overlooked: although he refers to the struggle for existence, the term actually means something much broader than this—*the struggle not only to exist, but also to reproduce.*

✳ The crux of the book comes with Chapter 4, entitled "Natural Selection." This was Darwin's big idea, an idea that is now largely taken as fact, but which in 1859 many found impossible to accept.

A POSITIVE SPIN

Chapter 3 of *The Origin* outlines what many of Darwin's opponents thought was an outrageous picture of the natural world, with living things locked in a brutal battle for survival. In the closing lines of the chapter, Darwin tried to put this bleak view in a better light:

"When we reflect on this struggle, we may console ourselves with the full belief, that the war of nature is not incessant, that no fear is felt, that death is generally prompt, and that the vigorous, the healthy, and the happy survive and multiply."

I'll have blue eyes and curly hair this time please

preserving favorable variations

NATURAL SELECTION

✱ Given that variations occur in living things, what happens when their owners compete in the "great and complex battle of life?" The answer, according to Darwin, was that favorable variations are preserved, while harmful ones are rejected and ultimately destroyed.

WEEDED OUT

✱ Darwin called this process natural selection. Although his phrase has stuck, it was not an entirely happy choice. The word "selection" is easily misconstrued. It suggests a deliberate act—something that involves a conscious selector—whereas what Darwin was describing was a process that involves no deliberate choices.

✱ With the consequences of natural selection, Darwin was on firmer ground. He listed a range of adaptations that he attributed to the gradual accumulation of useful variations. He pointed out that, unlike humans, nature bases its selection on the whole machinery of life. It weighs up the adaptive value of every single feature in living things, and in this process nothing gets overlooked.

CHOOSE YOUR PARTNER

✱ Darwin did identify one form of natural selection that can involve an element of choice. Called SEXUAL SELECTION, it accentuates any feature found in one sex—usually males—which tends to increase their chances of breeding successfully.

✱ Some of these features, such as the antlers of male deer, have an obvious use in fending off rival males. However, *in species where females actively choose mates, anything that attracts a female—no matter how useless it seems to be—works to a male's advantage*. For example, the flamboyant plumage of a male bird of paradise does not directly help it to survive. But if brilliant plumage proves effective in attracting more females, there will be a selective advantage in having it. As a result, selection will favor the most colorful males.

sheesh, I can hardly lift these!

some characteristics favor an individual's chances of finding a mate

IDEAS ABOUT INSTINCT

* The remaining chapters of *The Origin* deal with a range of topics that had occupied Darwin since his days aboard the *Beagle*. These include the evolution of physical adaptations, and also subjects as diverse as the origin of variation, instinct, fossils, classification, and the distribution of living things.

CHANCE AND CHANGE

Chapter 5 of *The Origin*, entitled "Laws of Variation," has not fared well against the test of time. Darwin began it by denying that variations are due to chance, although he admitted that "our ignorance of the laws of variation is profound." Today, biologists believe that variation is produced by random genetic change (see page 101).

let's see, what time is it?

Darwin was interested in how seeds travel

DOES INSTINCT EVOLVE?

* Darwin was keen to show that any inherited characteristic could be shaped by natural selection. Writing about animal behavior, he said **"I can see no difficulty in natural selection preserving and continually accumulating variations of instinct to any extent that may be profitable. It is thus, as I believe, that all the most complex and wonderful instincts have originated."**

* Darwin backed this up by examining some examples of instinct. For instance, many species of cuckoo occasionally lay eggs in other birds' nests—and this, Darwin argued, may have been how the ancestors of the Eurasian cuckoo once behaved. But, if one of these ancestral birds had an increased tendency to foist its eggs on other birds, it might well have produced more young. As a result, this pattern of behavior, initially an aberration, would become the norm.

MAKING A MOVE

✱ Darwin was particularly interested in the geographical distribution of living things. How likely was it, for example, that seeds could spread by floating across the sea? Darwin investigated by immersing seeds in sea water. **"To my surprise, I found that out of 87 kinds, 64 germinated after an immersion of 28**

seeds can travel across the sea

days, and a few survived an immersion of 137 days..." By checking the speeds of ocean currents, he concluded that 14 percent of plant seeds would survive a sea journey of nearly a thousand miles. If they were then cast ashore, they would stand a good chance of germinating.

Eurasian cuckoo

Someone else's nest

Most birds "put all their eggs in one basket," by laying them in a single nest. As a result, if their nest is destroyed, they lose the entire clutch. The Eurasian cuckoo, on the other hand, spreads the risk by laying up to 25 eggs, each in a different "foster" nest. As a result, more of its eggs are likely to survive. This increased survival rate provides a strong selective advantage for the cuckoo's sneaky behavior.

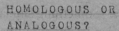

boing... boing...

mammals move around in a variety of ways

HOMOLOGY

* In his chapter on classification, Darwin discussed homologous structures — ones that share the same underlying plan despite being used in very different ways. Today, as in Darwin's time, HOMOLOGY provides one of the most powerful pieces of evidence for evolution.

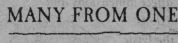

centipede

HOMOLOGOUS OR ANALOGOUS?

HOMOLOGOUS STRUCTURES — such as mammal limbs — are ones that share the same evolutionary origin. ANALOGOUS STRUCTURES — the wings of bats and birds, for example — are ones that have the same function, but which have evolved in very different ways. Analogous structures are produced by CONVERGENT EVOLUTION (see page 140), a form of evolution that creates similar adaptations in species sharing similar lifestyles.

MANY FROM ONE

* Darwin explored homology by looking at the limbs of mammals. As a group, mammals move about in a huge variety of ways, such as running, swimming, and flying. Their limbs vary enormously in size and shape, and each type seems purpose-built for a particular form of movement. Yet, *under the skin, mammalian limbs are all constructed on the same basic plan, using the same set of bones.*

* Something similar can be seen in the mouthparts of insects. Moths have long tubular tongues that can be conveniently coiled up before takeoff. Bugs have sharp piercing mouthparts that fold away when not in use, while beetles often have large and powerful jaws. But *despite their immense diversity, these too are built from just one set of parts.*

EXPLANATIONS

* Many 19th-century biologists asserted that these underlying similarities were simply the way that the Creator had decided to design things. Darwin, however, insisted that homology was *a product of natural selection. By making small but cumulative changes, natural selection can transform a leg into a wing or a flipper. But because natural selection cannot make sudden switches in body parts, these changes only affect dimensions and relative proportions. Underneath, the basic plan of the limb remains unchanged.*

I'm good at the digging but not at the running

Darwin on homology

"What can be more curious than that the hand of a man, formed for grasping, that of a mole for digging, the leg of a horse, the paddle of a porpoise, and the wing of a bat, should all be constructed on the same pattern, and should include the same bones, in the same relative positions?"

the compound eyes
of the housefly

ON SECOND THOUGHT...

Darwin constantly revised and updated the text of *The Origin*, with the result that it went through six different editions in his lifetime. Some of the changes helped to clarify the original version, but others show Darwin modifying his views in the face of criticism. The final edition, published in 1872, contains an additional chapter dealing with objections to the idea of natural selection.

the complex eyes
of the chameleon

OBSTACLES AND OBJECTIONS

* Darwin was one of his own toughest critics, and in *The Origin of Species*, he did not duck the many difficulties posed by his theories. One of the most intractable was explaining how complex organs—such as eyes—could ever have evolved by a series of extremely small steps.

EYES DOWN

* *Even Darwin had difficulty coming to terms with the idea that something as elaborate as an eye could be created by natural selection.* But logic says that this must be so. Throughout the animal kingdom, eyes vary enormously. The simplest eyes merely enable an animal to distinguish between light and dark, while the most complex provide detailed images. Between these are a vast number of intermediate forms, often built on similar underlying plans. *If these can exist, Darwin argued, it must be possible for elaborate eyes to evolve from simpler ones—an argument that holds for other "organs of extreme perfection" as well.*

eye

I'm off to live on the land

SWAPPING ROLES

✱ Another objection Darwin tackled was that of transitional lifestyles. *How could an aquatic animal slowly take up life on land, or a terrestrial mammal gradually become a bat?* Halfway through this process, the result would be something so mixed up it would have no chance of survival.

✱ Darwin countered these objections by showing that "halfway house" animals do exist in nature, and they manage to survive quite successfully. Frigate birds, for example, soar high over the sea, robbing food from other birds. Although they never land on the water if they can possibly avoid it, they still have webbed feet, indicating that their ancestors must have been water birds. Dippers, diving birds, are evolving in the opposite direction. They have slender toes built for hopping on land, but feed by plunging into streams.

Candid confession

A year after *The Origin* was published, Darwin admitted to the American botanist Asa Gray that "the eye to this day gives me a cold shudder." Nearly a century and a half later, opponents of natural selection—and evolution itself—often use the complexity of eyes to argue that some organs cannot have arisen by chance.

Asa Gray

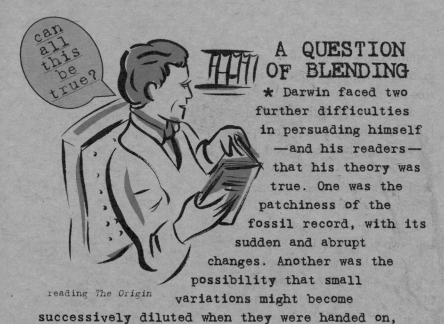

can all this be true?

reading *The Origin*

A QUESTION OF BLENDING

✱ Darwin faced two further difficulties in persuading himself —and his readers— that his theory was true. One was the patchiness of the fossil record, with its sudden and abrupt changes. Another was the possibility that small variations might become successively diluted when they were handed on, until they eventually had no effect at all.

Beating a retreat

In the final edition of *The Origin*, Darwin was forced to turn to Lamarckism (see page 48) to deal with the question of useful variations being diluted out of existence. He wrote that, although evolution is "effected chiefly through the natural selection of numerous successive, slight, favourable variations," it is "aided in an important manner by the inherited effects of the use and disuse of parts."

COMING CLEAN

✱ In *The Origin of Species*, Darwin devoted a whole chapter to what he called the "imperfection of the fossil record." He admitted that, according to his theory, geological strata should be packed full of fossils showing every transitional stage in evolution. But the reality is quite different. Instead of proceeding smoothly, evolution often seems to make sudden jumps. Darwin was firmly

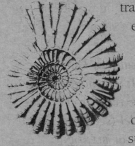

an ammonite

archaeopteryx

committed to the idea of gradual change, but conceded that this was a problem of the "gravest nature." He dealt with it by arguing that, because so many intervening species have left no trace, the fossil record is far from complete.

✱ Fortunately for Darwin, one very important transitional fossil—a primitive bird called *Archaeopteryx*—came to light only two years after *The Origin* was published. During the years that followed, other gaps in the fossil record were slowly filled, giving an added boost to Darwin's "gradualist" view.

FEATURES THAT FIZZLE OUT

✱ The problem of new features being <u>DILUTED</u> was first raised in 1867. A reader of *The Origin* pointed out that the effect of any useful variation would be halved each time its owner bred, unless its partner had the same variation as well. Over just a handful of generations, this halving would dilute the feature to such an extent that it would eventually disappear.

✱ *This was a problem that Darwin never managed to resolve. It was not until the 20th century, with the rediscovery of the work of Gregor Mendel (see page 98), that the idea of <u>DILUTION</u>—or <u>BLENDING INHERITANCE</u>—was finally overturned.*

DILUTE ME!

EARLY BIRD

During 1861 a fossilized <u>MISSING LINK</u> between reptiles and birds was found in a limestone quarry in Bavaria. Named *Archaeopteryx lithographica*, it showed a mixture of reptilian and avian features, including teeth, clawed fingers, and feathers. From Darwin's point of view, the discovery of the *Archaeopteryx* fossil could hardly have been better timed. It helped to endorse his view that the fossil record was not only incomplete but poorly explored, and that many intermediate life forms still awaited human discovery.

"THE MOST DANGEROUS MAN IN ENGLAND"

✱ The closing paragraph of *The Origin of Species* ends with a poetic flourish, describing the grandeur in the view that life, in all its extraordinary and rich diversity, is an evolving product of nature's laws. But, however carefully Darwin may have chosen his words, his conclusions contained in *The Origin* were bound to shock. Almost as soon as *The Origin* went on sale, the reverberations were underway.

I told you not to have that face lift

GOING UNDERGROUND

✱ By the time he had completed *The Origin of Species*, Darwin's health was in a precarious state. Things were not improved by his rapidly growing fame—or, as many people saw it, notoriety. He was often the subject of jokes, caricatures, and scathing articles in the press; and on one of his increasingly rare visits to London, an anonymous clergyman is reputed to have pointed him out to his companions as the most dangerous man in England.

✱ *Opinion about* The Origin *was divided, and the balance was not in Darwin's favor. Church leaders, in particular, were unhappy about the idea of evolution and were even more incensed by a concept of evolution that had no divinely guided purpose.*

✱ Darwin repeatedly referred to "the Creator" in *The Origin*, suggesting that he still believed in God. However, Darwin's god—if he still existed—was very different from the omnipotent God of the Christian church.

Darwin was not the Church's favorite person.

THE MONKEY CONNECTION

✱ A second issue sparked off even greater outrage and ridicule—the evolution of human beings. Darwin had said next to nothing about the place of humans in his evolutionary theory, writing simply that "light will be thrown on the origin of man and his history." *However, the implication was clear: humans had evolved from other animals—and the animals he had in mind, his readers concluded, were apes.*

✱ Arguments about God's omnipotence may have been above the heads of many members of the public, but this shocking suggestion was not. For the Victorian equivalent of today's tabloid press, Darwin and apes became inseparable.

Primate parade

In 1861—two years after *The Origin* was published—the French explorer Paul Belloni du Chaillu (1835–1903) staged an exhibition of stuffed gorillas in London. It was the first time gorillas, live or dead, had been seen outside Africa, and it generated great excitement. Many visitors to the exhibition recoiled at the idea that humans could have any biological connection with such "brutish" creatures.

I don't want to evolve into a human

evolving from apes

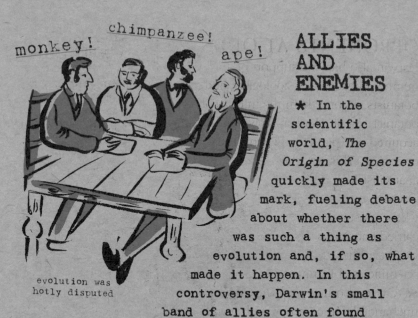

monkey! chimpanzee! ape!

evolution was hotly disputed

ALLIES AND ENEMIES

✱ In the scientific world, *The Origin of Species* quickly made its mark, fueling debate about whether there was such a thing as evolution and, if so, what made it happen. In this controversy, Darwin's small band of allies often found themselves at loggerheads with some of the most experienced and respected figures of the day.

The X club

Several of Darwin's supporters—including Thomas Huxley—were members of the "X Club," an informal group of leading British scientists. Working behind the scenes, X-Club members helped make sure that Darwin's work got a receptive hearing.

DARWIN'S BULLDOG

✱ Darwin himself did not have the temperament to throw himself into the fray, but he was hugely helped by a man who did—the biologist and lecturer **Thomas Huxley**, then in his early thirties. Darwin had sent a copy of *The Origin* to Huxley, who, after reading it, is said to have exclaimed *"How extremely stupid not to have thought of that!"* By a stroke of luck, Huxley was asked to review *The Origin* for *The Times*, which guaranteed it a favorable reception, and he continued to be one of Darwin's most important allies.

OPPONENTS AT ODDS

Darwin also had the support of a number of geologists, zoologists, and botanists, among them the influential botanist Joseph Hooker. However, he acquired one particularly dangerous opponent. This was the anatomist Richard Owen, a man whose prestige carried enormous weight in the scientific establishment.

★ Owen also reviewed *The Origin*, uncharitably. In common with many of his contemporaries, Owen did not dismiss the idea of evolution, but he could not accept the notion that it was driven by natural selection, rather than divine guidance. *He predicted that, within a decade, Darwin's work would be forgotten.*

★ Owen was not an easy man to get along with, and it is likely that his hostility was partly spurred on by professional rivalry. In contrast to Darwin, he failed to rally a group of like-minded people to his side. As a result, Darwin's ideas faced a disunited opposition and began to make headway. *Had Owen been less abrasive, things might have turned out differently, and Darwin's theory might have been smothered at birth.*

Richard Owen

Thomas Henry Huxley (1825–95) initially studied medicine, then joined the Royal Navy as a ship's surgeon. While in the navy, he took part in a four-year survey of the Australian coast, studying sea animals—particularly jellyfish—during off-duty hours. Biology became his passion, and in 1850 he gave up naval life to become a lecturer and later professor of natural history in London. Until 1859 he did not believe in evolution, but reading *The Origin of Species* changed his mind. Huxley's detailed knowledge of anatomy helped him to provide convincing evidence for evolution—although, like many of Darwin's supporters, he questioned the emphasis on natural selection.

BISHOP AND BULLDOG

* In the summer of 1860, during a meeting of the British Association for the Advancement of Science, a famous clash took place between Thomas Huxley— "Darwin's bulldog"—and Bishop Samuel Wilberforce. Huxley floored his opponent, scoring the first significant victory in the Darwinian campaign.

Genesis is true!

Captain Fitzroy of the *Beagle* disagreed with Darwin

SLUR ON THE FAMILY

* The historic encounter took place in Oxford before a large audience that included scientists, clergymen, university faculty, and students. No written accounts of it survive, but according to one eyewitness, the bishop—nicknamed "Soapy Sam" for his eloquence—was clearly eager for a fight. He talked "**with inimitable spirit, emptiness and unfairness**" and, after assuring the audience that there was nothing in the idea of evolution, delivered what was meant to be a knockout punch. *Turning to Huxley with a smile, he asked whether it was through his grandfather's or grandmother's side that he claimed descent from a monkey.*

Admiral Robert Fitzroy—formerly Captain Fitzroy of the *Beagle*—was also present at the Oxford debate and was firmly on the Bishop's side. He told the audience how he had often censured Darwin for airing views that conflicted with the book of Genesis.

✱ Huxley replied to good effect. *He defended Darwin's views and then added a stinging rebuke. While he would not be ashamed to have an ape for a grandfather, he would feel shame at being descended from someone who descended to cheap jibes to prove his point.* In reaction to Huxley's retort, cheering broke out, people jumped to their feet, and a lady is said to have fainted.

RUNNING A RETREAT

✱ The Oxford meeting proved to be an important milestone. Having been wrong-footed, the anti-evolutionists never regained the initiative, and from 1860 on, evolution gradually became accepted as a fact. But, despite Darwin's best efforts, it was still only a partial victory: *the mechanism behind evolution proved a much more difficult issue.*

Samuel Wilberforce

"Soapy Sam" Wilberforce (1805–73) was the son of William Wilberforce, the British philanthropist who campaigned for the abolition of slavery. A gifted mathematician and man of considerable charm, he was implacably opposed to the idea of evolution and made it known that he would use the Oxford meeting to "smash Darwin." Wilberforce was rumored to have been primed before the meeting by Robert Owen. Nevertheless, he made several errors during the course of the debate–which suggests that he had not fully grasped Darwin's arguments.

Huxley and Wilberforce arguing

America, the land of new ideas

DARWIN ABROAD

* Although *The Origin of Species* was first published in Britain, Darwin's ideas soon began to spread abroad. The reaction they provoked varied from country to country, illustrating how the course of science can be influenced by the prominent figures of the day.

CROSSING THE POND

* In the United States—a young country where new ideas might have been expected to take hold quickly—Darwinism had a mixed reception. The distinguished botanist **Asa Gray (1810–88)**, an old friend of Darwin and professor of natural history at Harvard, was a leading advocate; and further support was supplied by **Othniel Marsh (1831–99)**, America's leading dinosaur-hunter. Opposing them was a man with an even greater reputation, the Swiss-born naturalist **Louis Agassiz (1807–73)**. Agassiz not only rejected the idea of natural selection: he dismissed the idea of evolution altogether.

Louis Agassiz

ICE AGES AND EVOLUTION

Louis Agassiz is chiefly remembered for proposing the existence of ice ages. He was also a notable paleontologist, and published a book that described more than 1,500 species of extinct fossil fish. His extensive knowledge might have provided valuable evidence for evolution, but Agassiz maintained that such fossil species had been individually created.

* In the closing decades of the 19th century, most American scientists accepted that evolution occurred. But the concept of natural selection had a much rougher ride, and Americans tended to view it with even greater suspicion than their British counterparts.

LONDON CALLING

* Darwin's ideas initially failed to make much impact in France, but in Germany they were seized on with enthusiasm, particularly by a new generation of scientists anxious to make their mark. One of Darwin's most energetic converts was **Ernst Haeckel (1834–1919)**, a much-traveled naturalist and marine biologist who was professor of zoology at the University of Jena. *Haeckel made no attempt to patch over the differences between Darwinism and the teachings of the Church, arguing instead that natural selection and evolution did away with the need for God.* At a time of widespread discontent with established authority, it was a potent message that found a ready audience.

* *In his zeal, Haeckel went much farther than Darwin himself. He was the first person to popularize the "evolutionary tree" as a way of showing how species might have split and evolved, developing into the groups recognizable today.*

we love Darwin done German style

Darwin's ideas traveled to Germany

ADVOCATES OF PROGRESS

Unlike Darwin, Ernst Haeckel was fully wedded to the belief that evolution inevitably led to increased complexity in living things. Like Thomas Huxley in Britain, he was responsible for creating the widespread impression that evolution is synonymous with progress, the ultimate result of progress being human beings.

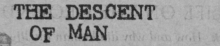

ape

THE DESCENT OF MAN

✱ In 1871 Darwin finally felt he could publicly broach the subject of human evolution. In *The Descent of Man*, he laid out the case for believing that humans and apes share a distant ancestor and that all human characteristics, no matter how unusual, have evolved in a series of gradual steps.

NEANDERTHAL MAN

In 1856, workers in Germany's Neanderthal (Neander valley) found a collection of bones in a limestone cave. The bones were clearly human, but the skull had a sloping forehead and large ridges over the eyes. The find provoked controversy,. with some scientists believing that it was a "missing link" between humans and apes, and others dismissing it as the remains of someone suffering from a bone disease. It turned out that neither side was right, but Neanderthal Man added fuel to the debate about human evolution.

NOT US, SURELY?

✱ In the 1870s, most scientists still saw humans as being separate from the rest of the animal world. Even among Darwin's closest supporters, the whole subject of human evolution was a highly contentious matter. Thomas Huxley firmly believed in gradual human evolution, but Alfred Wallace eventually rejected it, deciding that natural selection alone could not account for many human features.

✱ In *The Descent of Man* Darwin's main aim was to show that even the most remarkable human attributes—our intelligence and emotional expression—could have been produced by natural selection, allowing us to evolve from animal ancestors.

how long have I been here

LIFE ON THE GROUND

✱ *How and why did human intelligence arise? Darwin's view was that it was linked to a change in lifestyle.* The common ancestors of humans and apes were originally tree-dwellers, but they gradually took up life on the ground. Walking on two feet freed their hands to manipulate objects and, eventually, to make tools. This, Darwin argued, provided the springboard to intelligence,

Eyewitness account

In the 19th century, many people thought of humans as having appeared at a recent date in the Earth's history. One of the earliest pieces of evidence for human antiquity came in the 1860s, when the French paleontologist **Edouard Lartet (1801–71)** discovered a mammoth tooth with an engraving of a mammoth on it. The artist had clearly seen the prehistoric creature in the flesh.

fossil finds were very rare

because natural selection would then have favored the increased brain size needed for manual dexterity.

✱ Darwin's views about the evolution of intelligence met a very mixed response, but they echo ideas widely held today.

DARWINISM IN DECLINE

* At the time of Charles Darwin's death, in 1882, the battle to convince scientists about evolution was largely won. Darwin himself had achieved international fame - and he was buried in Westminster Abbey, a remarkable tribute to a man who had so often been denounced by the Church. But, paradoxically, one of Darwin's key ideas was still under attack; and in the years that followed, it seemed that a large part of his achievement would not survive.

IS SELECTION CREATIVE?

some variations are beneficial... others aren't

* The core of the problem was natural selection. Darwin's supporters, and many of his opponents, had no difficulty with the idea that nature might keep species "pure" by weeding out individuals that were infirm or badly adapted. *But the idea that natural selection could act as a positive force—encouraging the development of useful characteristics and creating new species—proved much harder for many people to accept.*

* According to the Lamarckist view of evolution (see page 48), variations in living things arise in response to the demands of everyday life. These variations

if I had another set of hands, I'd mow the lawn

beneficial variations?

are always positive. *The evolutionary mechanism that Darwin proposed works in quite a different way. Here, variations occur spontaneously and at random: some are beneficial while others are not, and natural selection creates new structures and lifestyles in an entirely piecemeal way.*

★ Today, this apparently haphazard process is generally accepted as the engine of evolution. But, *during the years after Darwin's death, the idea of natural selection as the driving force behind evolution attracted less and less support.*

★ The chief reason for the decline of Darwinism was that Darwin's work was based solely on observation. In his day, the mechanisms that lie behind variation and inheritance were unknown. *Until the 20th century, when those mechanisms started to come to light, there was a real possibility that the idea of natural selection would eventually become extinct.*

A NIGHTMARE OF WASTE AND DEATH

The English author Samuel Butler (1835–1902), who wrote the Utopian satire *Erewhon* (1872), proved to be one of Darwin's most scathing critics. Initially, he was an enthusiastic supporter of Darwinism, but he became opposed to the idea of natural selection as the driving force behind evolution, preferring his own brand of Lamarckism. He openly accused Darwin of having belittled Lamarck and described natural selection as a "nightmare of waste and death."

Samuel Butler

CHAPTER 3

EVOLVING INTO OBLIVION

Examples of "NON-ADAPTIVE" STRUCTURES, once thought to be produced by orthogenesis, included the stabbing canine teeth of saber-toothed cats. According to the theory's supporters, these teeth grew slightly longer with each successive generation, until their unfortunate owners were no longer able to close their jaws, condemning them to extinction through starvation. *Darwinian theory maintains that things could never have got this far, because natural selection would have halted tooth enlargement the moment it began to have a negative impact on survival.*

INTO THE 20TH CENTURY

* In rejecting natural selection, turn-of-the-century biologists were faced with the problem of finding an alternative mechanism behind evolutionary change. Lamarckism began to come back into fashion, together with a new evolutionary driving force called ORTHOGENESIS.

These antlers are getting ridiculous!

Irish elk

CAREERING AHEAD

* *Orthogenesis means development in a straight line. It was originally devised to explain apparently nonadaptive structures that are sometimes found in living things.* For example, fossils unearthed in Victorian times showed that the Irish elk, which died out after the last ice age, had colossal antlers weighing more than 110 pounds—about a seventh of the

saber-toothed tiger

weight of the entire animal. Laden with these huge weapons, a male deer would have found it difficult to move its head quickly to protect itself.

✱ *Many biologists found it impossible to see how natural selection could have produced features like these. Orthogenesis seemed to provide an answer.*

EVOLUTIONARY OVERSHOOT

✱ *According to the theory of orthogenesis, physical structures develop as the result of straight-line evolutionary change: once evolution has embarked on a particular course, it develops its own momentum.* In some cases, this process produces useful results, such as human intelligence; in others, it eventually burdens a species with a major handicap.

✱ *Orthogenesis attracted followers for several decades, but the idea had some serious flaws. No one could identify the force that drove evolution forward, and fossils contradicted the idea of smooth straight-line change.* Equally significantly, no one could prove that "overdeveloped" structures were actually a disadvantage. This opened up the possibility that natural selection might have produced them after all.

Help or hindrance?

The so-called Irish elk in fact lived throughout northern Europe and Asia. It was the largest known member of the deer family and was abundant during the height of the last ice age, about 20,000 years ago. Like all deer, its antlers were formed of bone and, despite their immense size, they were grown and shed every year.

"we support orthogenesis"

orthogenesis attracted followers for many decades

GREGOR MENDEL

* By a curious irony, the first glimmers of light on the ve^xed question of variation and heredity came while Darwin was still alive. Unknown to him, an Austrian monk named Gregor Mendel had conducted a series of experiments on pea plants and had made a remarkable discovery: instead of blending together, inherited characteristics are passed down the generations intact.

Mendel and his pea plants

Why peas?

Mendel used peas because they have a number of characteristics that show clear-cut differences from plant to plant. They are also easy to grow, and they have the advantage that they can pollinate themselves.

By using several generations of self-pollinated plants, Mendel was able to establish "pure-bred" lines for his research.

MIX AND MATCH

* Mendel carried out his research in a monastery garden at Brünn (now Brno in the Czech Republic). During the mid-1850s, he selected pea plants that had distinctive and contrasting features—such as red or white flowers, and green or yellow seeds—and then interbred them to see the results. *According to the accepted ideas of the time, these inherited characteristics should have blended.* For example, crossing red-flowered and white-flowered parents should have produced a new generation of plants with flowers that were neither red nor white, but

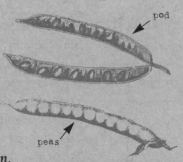

pod

peas

somewhere in between. *However, Mendel found out that the reverse was true. The flowers in the new generation were either red or white— just like those of their parents. More intriguing still, he found that in some situations a characteristic could apparently disappear in one generation, only to reappear in a later one.*

✱ Mendel carried out thousands of cross-breeding experiments and analyzed the results. *The statistics showed that characteristics such as red or white flower color were passed on according to simple rules and were inherited in a ratio of 3:1.*

OVERLOOKED EVIDENCE

✱ *Mendel's experiments might have rocked 19th-century science. His work, however, was largely ignored.* Mendel presented his findings to the Natural Science Society in Brünn in 1865, but his experiments and pioneering use of statistics attracted little interest. One leading scientist who did read his work, the German botanist **Carl von Nägeli**, told him that his results were not significant and he should have experimented with more plants. *By then, Mendel had already cross-bred more than 20,000. His work—and its implications for evolutionary theory—lay dormant for another 35 years.*

Gregor Mendel (1822–84) entered the monastery at Brünn at the age of 21. He spent two years studying science and mathematics at the University of Vienna, before returning to the monastery and beginning his plant-breeding experiments. From a statistical point of view, some of Mendel's results were almost too good to be true. In one experiment, for example, he checked 8,023 plants, and found a ratio for seed color of almost exactly 3:1. *Results like these have made statisticians suggest that Mendel, or perhaps an assistant, "massaged" the figures to achieve the outcome he wanted.*

VARIATION AND INHERITANCE

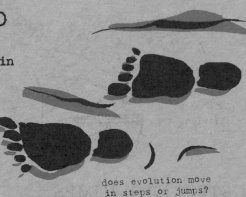

* By 1900, interest in heredity had grown, and other scientists were unwittingly following in Mendel's footsteps. During the course of their research, Mendel's work was rediscovered, and the laws that govern heredity at last became known. Mendel's observations ultimately vindicated the idea of natural selection, but only after lengthy arguments about how they should be applied.

does evolution move in steps or jumps?

INVENTING GENES

In 1909, "hereditary elements" gave way to "genes"—a more convenient term, coined by the Danish botanist Wilhelm Johannsen (1857–1927).

KEY WORDS

GENE:
a unit of heredity

MUTATION:
in its original sense, any major and abrupt change that can be passed on by heredity; in its modern sense, the word means any change involving chromosomes or individual genes (see page 106)

THE END OF BLENDING

* One immediate difficulty solved by Mendel's work was the problem of dilution, which was first raised in 1867 (see page 83). *Mendel had discovered that characteristics are controlled by <u>FACTORS</u> he called "<u>HEREDITARY ELEMENTS</u>," which cannot be blended together. If one of these factors produces a useful characteristic, there is nothing to stop natural selection from favoring it through successive generations, making the characteristic more widespread.*

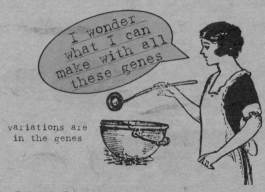

I wonder what I can make with all these genes

variations are in the genes

STEPS AND JUMPS

✱ A more difficult problem was the nature of variation. Darwin believed in CONTINUOUS VARIATION—*variation consisting of an almost infinite array of tiny differences between individuals.* But by 1900 many biologists had become convinced that variation was actually DISCONTINUOUS—*with occasional abrupt changes, called* MUTATIONS, *providing the raw material of evolution.*

✱ When Mendel's work was rediscovered, the "MUTATIONISTS" seized on it as evidence that they were right because it seemed to show how sudden and abrupt variations might be handed on. For their part, the Darwinian "SELECTIONISTS" dismissed mutations as being of little importance in evolution as a whole.

✱ By the 1930s, it was clear that neither side had a monopoly on the truth. *Mendel's "hereditary elements"—now known as* GENES—*turned out to behave in ways that accounted for variations large and small.*

JUMPING TO CONCLUSIONS

The Dutch botanist **Hugo de Vries (1848–1935)** was one of three scientists who in 1900 independently uncovered Mendel's work. He carried out a long series of breeding experiments with evening primroses, after noticing a group of plants that seemed to show large and discontinuous variations. *His book* **Mutation Theory,** *which was published in 1901, advocated the idea that evolution is driven by sudden jumps, which transform one species into another.* De Vries' work turned out to be built on shaky foundations, because it was later shown that most of his evening primrose "mutations" were the result of recombination of existing characteristics, rather than the formation of completely new ones.

Hugo de Vries

101

genes, like ants, work in teams

DARWINISM REVIVED

* How can genes produce variation that is smooth, as well as variation that is discontinuous? This paradox puzzled Mendel, and it was not one he was able to resolve. In the opening years of the 20th century, geneticists discovered the answer: characteristics are often controlled not by single genes, but by a number of genes working together.

LINKAGE

Mendel's second law is now known to be only partly true, because genes situated close together on chromosomes are often passed on together. This phenomenon is known as LINKAGE. Mendel's pea experiments avoided the complication of linkage because, by sheer luck, most of the features he studied were on different chromosomes.

Mendel was puzzled by continuous variation

MENDEL'S LAWS

* As luck would have it, when Mendel studied heredity in peas, he chose seven distinct characteristics that are each controlled by a single gene. From his breeding experiments, he was able to deduce his two laws of heredity. *The first law states that the characteristics of living things are controlled by paired "factors" (genes), with one of every pair coming from each parent. A single gene may have alternative forms, known as ALLELES; and, where this is the case, one allele in a pair* is often <u>DOMINANT</u>, *masking the effect of the other, which is* <u>RECESSIVE</u>.

* *Mendel's second law states that genes are dealt out independently during reproduction, instead of being handed on in groups. Offspring inherit their parents' genes—but, because of this system of independent assortment, they have a new and unique combination of alleles. This is one source of the variation on which natural selection can act.*

* But what about characteristics like human height, which do not vary in a series of discontinuous steps? Here, many more genes are in action, reflecting the fact that growth is a highly complex process. *These* <u>POLYGENIC CHARACTERISTICS</u> *are much more difficult to study, but the genes still follow the same underlying rules. Instead of producing offspring that vary in jumps,*

tall, medium, and small

they produce a range of variation clustered around an average. On a graph, this statistical spread, which is known as a <u>NORMAL DISTRIBUTION</u>, shows up as a bell-shaped curve.

SPREADING OUT

One of the first studies of polygenic inheritance was carried out by the American geneticist Edward M. East in the early 1900s. He selected corn plants with long or short ears, and then cross-bred them. The result was a generation of plants with a wide variety of ear size, clustered about a size roughly intermediate to the two parents. In the 19th century, this would have been seized on as an example of "blending inheritance" (see page 83), but East recognized that it was produced by the combined effects of many genes, resulting in almost continuous variation.

Support for Darwin

Genes interact in many ways. In the 1920s and 1930s, it became clear that polygenic inheritance meant that Mendelian heredity did not rule out Darwin's ideas on variation and natural selection. Instead, it bore them out.

REDATING THE EARTH

* With the birth of genetics, the theory of evolution by natural selection underwent a slow but steady renaissance. At the same time—thanks to discoveries in a completely unrelated field— one of Darwin's greatest problems was finally solved.

when was the Earth's birthday?

HALF-LIFE
Radioactive elements give off energy when their atoms decay into atoms of other elements—a process that happens at a precise and invariable rate. This is measured by an element's HALF-LIFE—the time it takes for exactly half a sample's atoms to decay. Half-lives vary from a fraction of a second to tens of billions of years.

RUNNING SHORT OF TIME

* *In the 1860s the Scottish physicist William Thomson (1824–1907), who later became Lord Kelvin, carried out some detailed calculations about the age of the Earth.* Assuming that the Earth was initially a ball of molten rock, he worked out how long it would take the planet to cool sufficiently for a solid crust to form. Kelvin was one of the foremost scientists of his day and a world expert in thermodynamics. As a result, the figure he came up with—between 20 and 100 million years—had great authority.

* For Darwin, Kelvin's calculations had disastrous implications. While 100 million years is an immense span by human standards, Darwin realized that it was not nearly long enough for life to have evolved in the way he had proposed. Privately, he doubted Kelvin's conclusions, but there was little he could do to dispute them.

THE RADIOACTIVE EARTH

* In 1896—14 years after Darwin's death—a chance discovery led to a complete rethinking of the Earth's age. While he was investigating fluorescent substances, the French physicist **Antoine-Henri Becquerel (1852–1908)** accidentally discovered radiation. In 1901 his compatriot **Pierre Curie (1859–1906)** measured the amount of energy given off by the radioactive element radium. He found that it was not only extremely high, but that it declined at an almost imperceptible rate.

* Radioactive elements make up only a small proportion of the Earth's total mass, but the energy they release is enormous. **Charles Darwin's son George (1845–1913),** *who became an astronomer, was among the first to realize that this newly discovered source of energy invalidated Kelvin's calculations, making the Earth far older than was previously thought. The age generally accepted today—about 4.6 billion years—is long enough for complex life to have evolved through an almost infinite series of steps.*

RADIOACTIVE DATING

Uranium 238 is a naturally occurring ISOTOPE that decays to form lead 204. It has a half-life of 4.5 billion years. *Decay proceeds from the moment uranium is incorporated into rock—so by measuring the relative proportion of uranium 238 and lead 204 in a rock sample, a rock's age can be estimated.* This dating technique shows that the most ancient rocks on Earth formed about 4 billion years ago.

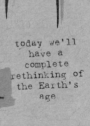

today we'll have a complete rethinking of the Earth's age

HARD AND SOFT HEREDITY

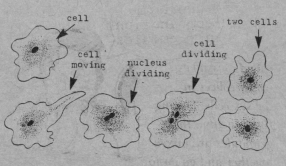

cell

cell moving

nucleus dividing

cell dividing

two cells

✱ In 1902 Walter Sutton, a young American biologist, correctly guessed that genes are carried on chromosomes, the threadlike structures found in the nuclei of living cells. This breakthrough provided new impetus to a long-running debate: can external forces influence heredity to make living things adapt?

CHROMOSOMES AND GENES

Most living things have only a few dozen chromosomes per cell—not nearly enough if each chromosome merely controlled a single characteristic. *In 1906 the British geneticist William Bateson (1861–1926) realized that each chromosome must carry a whole range of separate "factors" (i.e. genes) that control a wide variety of characteristics.*

THE IMMORTAL LINE

✱ *Most living things start life as a single cell. During early development the original cell divides, eventually producing a multitude of new cells that are specialized for different roles. <u>SOMATIC CELLS</u> ("BODY CELLS") form structures such as bones, muscles, and nerves, while <u>SEX CELLS (ALSO KNOWN AS "GERM CELLS")</u> are set aside for reproduction.*

✱ During the 19th century there was a lot of confusion about how sexual reproduction works. *Darwin believed in <u>PANGENESIS</u>. According to this theory, all parts of the body form tiny particles*

human embryo

I'm a handsome combination of gemmules

called GEMMULES, which collect in the sex cells. During reproduction, the gemmules are mixed together and go on to form the parts of the body that originally produced them.

✱ In the 1890s, the German biologist **August Weismann (1834–1914)** published an alternative theory. *Somatic cells, he believed, die with their owner, but sex cells form a continuous line, carrying hereditary material or "GERM PLASM" from one generation to the next.*

ONE-WAY STREET

✱ *Weismann's germ-plasm theory was largely correct, although genes—rather than cells—turned out to be the real players in the game.* His idea of a one-way link between germ cells and somatic cells ruled out "SOFT HEREDITY"—the Lamarckist notion that changes acquired during life can somehow affect sex cells and then be passed on.

✱ The link between chromosomes and genes, posited by Sutton, established a physical basis for "HARD HEREDITY," in keeping with Mendel's ideas. *Despite attempts to disprove it—described on the following pages—hard heredity is an integral part of genetics today.*

KEY WORDS

CHROMOSOME:
a threadlike structure that carries genes; each species has a characteristic number of chromosomes in its cells, but the chromosomes normally become visible only when they "condense," just before a cell divides

one missing tail

Crucial cuts

August Weismann challenged the idea of soft heredity by breeding 22 generations of mice, cutting off their tails at birth, and then searching for any young that developed stunted tails. He didn't find any, but could have avoided his grisly experiments: after centuries of circumcision, Jewish males are still born with foreskins—providing just the kind of evidence Weismann was seeking.

THE TALE OF THE MIDWIFE TOAD

your babies are safe with me

* One of the strangest episodes in the history of evolutionary research took place just after World War I. <u>Paul Kammerer</u> (1880-1926), an Austrian biologist, claimed to have discovered a case of "soft heredity" in an animal called the midwife toad. Ultimately, his evidence failed to stand up—and, in an atmosphere of distrust and suspicion, Kammerer shot himself.

IDEOLOGICAL IMPLICATIONS

Kammerer used his midwife toad experiments to suggest that humans could influence the course of evolution—an idea that was quickly taken up by the popular press. This notion fell on particularly fertile ground in the Soviet Union, where Marxist ideology was at odds with Darwinian natural selection.

REAPPEARING PADS

eggs

* The midwife toad is an unusual animal. Unlike most European toads, it mates on dry land, and the male collects the female's eggs and wraps them around his hind legs. The male carries the eggs for about a month, acting as a midwife until they are ready to hatch. He then makes his way to water, where the tadpoles are released.

* Midwife toads have several adaptations that suit them to a largely terrestrial life. *The one that interested Paul Kammerer was their lack of "nuptial pads"—the*

tadpole

rough swellings that other male toads have on their thumbs.

These pads help the males to grip the damp females when they mate.

✱ Kammerer raised several generations of midwife toads in captivity, keeping them more damp than they would have been in the wild. *He claimed that when the toads were treated in this way, the males developed nuptial pads, and they handed on this characteristic when they bred.*

THE END OF THE AFFAIR

✱ This apparent example of "soft heredity" roused a mixture of public interest and scientific scepticism. Other scientists failed to reproduce Kammerer's results; and the British geneticist William Bateson asked to see one of Kammerer's toads, so that he could verify the effect himself. Kammerer proved reluctant to part with any of his specimens, but in 1926 an independent examination was finally arranged. *It concluded that the nuptial pads had been produced by injections of Indian ink.*

✱ Kammerer insisted that he had not manufactured the pads himself, but a short while later took his own life. Throughout, he had maintained that his observations were genuine and that there was an innocent explanation for the ink. However, one result of his suicide was that the full facts were never revealed.

Kammerer and Koestler

The Hungarian-born writer Arthur Koestler (1905–83) followed an established literary tradition when he wrote a sympathetic account of Kammerer's experiments, called *The Case of the Midwife Toad*. Like other influential writers, such as the novelist Samuel Butler (see page 95) and the playwright George Bernard Shaw (1856–1950), Koestler believed that Lamarckism, or "soft heredity," must play some part in evolution. In his book, Koestler concluded that Kammerer's results were authentic.

Kammerer kept his toads damp and observed the results

Trofim Lysenko

EVOLUTION SOVIET STYLE

* Until the late 1950s, Lamarckism had one last stronghold: the Soviet Union. Under the leadership of the botanist Trofim Lysenko (1898-1976), Soviet policy-makers officially rejected both Mendelian genetics and Darwinian natural selection —with calamitous results.

Is natural selection a capitalist idea?

During the Lysenko era, Soviet political scientists denounced Darwinism and Mendelian genetics as the products of capitalist thinking. Some historians think that the parallels between natural selection and "laissez-faire" capitalism are not merely coincidental. Ideas in science are often influenced by the social and political climate of the time. According to this view, it is hardly surprising that Darwin's view of nature was shaped by a society in which the weakest often went to the wall.

SIDESTEPPING SCIENCE

* Lysenko introduced a procedure called VERNALIZATION, which involves subjecting seeds to low temperatures to make them germinate faster in spring. On the strength of this "discovery"—which was already

Lysenko believed that vernalizing grain altered its genetic makeup

known to agriculturalists in the West—Lysenko rose through the ranks of the scientific establishment.

✶ Lysenko believed that vernalizing grain altered its genetic makeup, allowing early germination to be inherited. He also claimed that he could make permanent improvements to grain varieties simply by growing the parent plants in better conditions. In 1948, the Communist Party declared that Mendelian genetics and natural selection were erroneous, allowing Lysenko to introduce his own brand of Lamarckism in their place.

COURTING DISASTER

✶ In the repressive atmosphere of the time, few geneticists dared to object to Lysenko's methods. Those that did lost their jobs and often ended up in prison. **Lysenko's techniques were applied in Soviet agriculture, in the expectation that productivity would climb. However, despite a huge investment in mechanization, the dramatic improvement in harvests failed to materialize.** To make matters even worse, the Soviet Union was denied the chance to benefit from crop improvements taking place in the West, which had been made by applying Mendelian principles.

✶ By the 1950s, Lysenko's scientific failings were becoming apparent, but he still managed to maintain his influence. He eventually resigned in 1965, having inflicted damage that took years to repair.

Nikolai Vavilov (1887–c.1943)

The Russian botanist Nikolai Vavilov was one of the most prominent victims of Lysenko's rise to power. After training in Russia and Britain, Vavilov became the director of the Soviet Institute of Plant Industry in 1920, and during the 1920s he traveled the world to collect potentially useful plants. However, Vavilov firmly believed in Mendelian genetics, and this proved to be his undoing. He was arrested and sentenced to death in 1940, and is believed to have died in a labor camp in 1942 or 1943.

collecting clues

REVEALING GENES

* For evolution to work, information has to be stored by genes in a form that can be handed on. But what exactly are genes made of? The answer to this question began to take shape with a chance observation made in the late 1920s, and it eventually led to one of the 20th century's greatest breakthroughs: the discovery of the structure of DNA.

STORING DNA

In most living things, DNA is stored in the nuclei of living cells. The only exceptions to this rule are bacteria, which do not have nuclei. Instead, their DNA is dispersed throughout their cells.

Nucleic acids

These were first discovered in 1869, but their role in cells was a mystery. By the 1940s it had been established that they are built of small units called NUCLEOTIDES, but their structure and function were still unknown.

MESSAGES FROM THE DEAD

* In 1928 **Fred Griffith**, a British microbiologist, noticed a curious thing. While working with a bacterium that has two distinct forms, he found that one form could be transformed into the other simply by mixing it with the other's dead cells. Griffith's findings remained an unexplained oddity until the early 1940s, when a group of scientists at the Rockefeller Institute decided to repeat his work. *They discovered that the transformation was inherited, and that it was triggered not by the dead cells themselves but by a substance inside them—DEOXYRIBONUCLEIC ACID, or DNA.*

THE DNA ENIGMA

✱ The rerun of Griffith's experiments came close to proving that DNA contains GENETIC INFORMATION. *New microscopic techniques, using ultraviolet light, revealed that DNA is packed up in chromosomes, and researchers also discovered that the amount of DNA in cells is constant, with sex cells having exactly half the amount of somatic cells.* Both these finding were exactly what would be expected if DNA and genes are one and the same.

✱ Once the link was beyond doubt, geneticists faced a problem. Previously, genes were suspected of being made of proteins—highly variable molecules composed of amino-acid units. *But chemical analysis showed that DNA is simpler than a protein, although its molecules are much longer.* Somehow, this relatively simple molecule had to store genetic information in a way that could be used, copied, and handed on. The question was, how?

Griffith's experiment

Griffith's experiments

Fred Griffith's discovery of transformation came while he was studying *Streptococcus pneumoniae*, a bacterium that causes pneumonia. One form of the bacterium has a smooth outer coat. The other form lacks this coat and has a rough surface; unlike the smooth form, it does not produce disease. Griffith found during his experiments that the rough form could be changed into the smooth form by contact with dead smooth-coated cells, transforming harmless bacteria into ones that were capable of producing disease.

how does DNA store genetic information?

SPIRAL OF LIFE

* The structure of DNA was established in 1953 by the American biochemist James Watson and the British

James Watson

physicist Francis Crick. They revealed a substance that is capable of preserving and duplicating infinitely varied sequences of information, using a chemical code common to all living things.

JUNK DNA

In a DNA strand, there are no physical markers that show where each gene begins and ends. Instead, genes are marked off by particular sequences of base-pairs. Genes are often separated by long sequences of bases that do not contain any identifiable genetic information. In humans, this so-called "JUNK DNA" makes up over 95 percent of each DNA strand.

Vital statistics

A single DNA molecule can be more than a millimeter long, and may contain over 5,000 genes. A remarkable list of chemical coincidences means that, no matter how the base-pair sequence is arranged, the molecule always keeps its shape.

the bases are A,G,T, and C

OPPOSITES ATTRACT

* A DNA molecule combines two apparent opposites: regularity and infinite variety. The regularity is provided by the twin backbones of the molecule, a pair of helical strands that wind around each other. *These strands are built of two alternating chemical units repeated millions of times. Because their structure is completely regular, the strands contain no information—just as two letters, endlessly repeated, convey no meaningful message.*

* The variety comes in the links that hold the two strands together. These are created by four chemical bases: ADENINE (A), THYMINE (T), GUANINE (G), and

have I got the right pairs?

the bases form pairs

CYTOSINE (C). Each link consists of a pair of bases; adenine always partners thymine, while guanine partners cytosine. This gives four possible base-pairs—AT, TA, GC, and CG—which can be arranged in any order. *The order that they follow forms a store of information which varies from one living thing to another. This store is duplicated and handed on when cells divide.*

USING INFORMATION

✱ Stored information, on its own, is not much use to living things. To be of value, it must be capable of being acted on, or EXPRESSED. Expression is a complex process, but its outcome is always the same. Genes—which are simply specific lengths of the information sequence—instruct cells to assemble proteins. Proteins are then used to construct cells, or to modify the way they work.

✱ *By controlling the formation of proteins, genes indirectly shape every inherited characteristic of living things.*

duplicating the original

ERRORS IN THE CODE

✱ Despite being a million times thinner than a human hair, DNA molecules are remarkably resilient. They have to be, because without reliable genetic instructions, living things could not survive. But occasionally, those instructions are accidentally altered, with unpredictable and sometimes far-reaching results.

don't alter the instructions

genetic instructions have to be reliable

Vanishing mutants

Mutations in sex cells are the only mutations that can be inherited, so they are also the only mutations that are subject to natural selection. Mutations in somatic or body cells (see page 106) cannot be handed on; they disappear with the death of their owner.

THREE-LETTER WORDS

✱ When a gene is put into action, its base sequence is first copied and then decoded—twin processes known as TRANSCRIPTION and TRANSLATION. In the translation stage, the gene's sequence of bases is "read" in groups of three. Each group, or CODON, instructs the cell to collect one of the 20 amino acids that cells use to make proteins. Once the amino acid is in position, it is then added to a growing string of amino acids to form the protein molecule.

✱ The GENETIC CODE that links bases to amino acids is extremely precise, and a single error can have large repercussions. If part of a DNA molecule is damaged—for example, by ultraviolet light from strong sunshine—this may alter the base sequence of a gene, producing an abnormal protein. This abnormality may be passed on when the DNA copies itself.

✱ *Alterations like these are known as mutations. This term was originally used to describe any abrupt change, but in modern genetics, it has acquired a much more precise meaning. Mutations are not only produced by small-scale damage to DNA. Sometimes, entire chromosomes may be lost or duplicated, causing a major change to the way cells work.*

DEALING NEW CARDS

✱ If DNA preserved its information with unerring precision, the only source of variation in living things would be the reshuffling of genes that occurs during sexual reproduction. New genes would never come into existence, so natural selection would merely alter the relative frequency of the genes that were already in existence.

✱ *Mutations provide a constant injection of new variation that can be put to the test. Because they are random, very few of them are beneficial—but without them, new adaptations could not evolve.*

RUNNING REPAIRS

Like all organic substances in living cells, DNA is subjected to a constant barrage of chemical activity from other substances around it. This creates damage which, if left untreated, could lead to a torrent of mutations. However, most forms of damage are quickly patched up by REPAIR ENZYMES—proteins that correct chemical abnormalities before they have a chance to interfere with genetic activity. Without this repair system, mutations would be so frequent that protein molecules would often be defective.

I always seem to be patching up these proteins

THE MODERN SYNTHESIS

hi, man

a modern Darwin

* Had Darwin lived in the 20th century, rather than the 19th, he would have found that being a well-read naturalist was not enough to keep abreast of evolutionary theory. Statistics and MOLECULAR BIOLOGY have helped to transform evolution from an abstract process into something that can be traced and analyzed in a quantitative way.

KEY WORDS

HOMOZYGOTES AND HETEROZYGOTES
Individuals with contrasting alleles of a particular gene are known as HETEROZYGOTES. Their recessive alleles represent a hidden reserve of genetic variation that can be exploited by natural selection if living conditions change. Individuals with identical alleles of a gene are known as HOMOZYGOTES.

THE NUMBER CRUNCHERS

* One of the first people to use statistics to study evolution was the British mathematician **Ronald Fisher (1890–1962)**. Working long before the development of computers, Fisher produced mathematical models that predicted the effects of natural selection. When he began his work in the 1920s, many biologists believed that Darwin's gradualist approach was at odds with Mendelian genetics. Fisher's results— for those who were able to follow his highly complex calculations—showed that natural selection could account for the smooth range of variation seen in living things.

fruit fly

* **Fisher's work helped to open up a new field of study—** POPULATION GENETICS. **It was fueled by a growing realization that evolution is not so much about individual organisms as about the genetic identity of whole populations. Evolution occurs whenever there is a change in the overall frequencies of the population's genes, even if that change has no visible effect.**

DARWIN UP TO DATE

* In 1942 **Julian Huxley**, grandson of Darwin's ally Thomas Huxley, published a review of evolutionary theory entitled *Evolution: The Modern Synthesis*. It marked the complete rehabilitation of Darwinism, following years of disagreement about the validity of natural selection. *Supported by evidence that was unavailable to Darwin himself, it ushered in the era of neo-Darwinism, an up-to-date account of evolution by natural selection.*

* Since the structure of DNA was established in 1953, the science of molecular biology has developed with bewildering and dramatic speed. Successive discoveries about the nature of genes and the complex ways they interact seem to have validated Darwin's belief in the power of natural selection.

HIDDEN VARIATION

There are two forms of variation in living things: variation that is outwardly expressed, and variation that is hidden away in the form of recessive alleles (see page 102). In the 1920s the Russian naturalist Sergei Chetverikov carried out a series of experiments that revealed some of this HIDDEN VARIATION. He crossed wild fruit-flies with pure strains raised in the laboratory and identified new features that appeared in their offspring. These features were produced by recessive alleles that are normally masked in the wild.

evolution is about whole populations

119

CHAPTER 4

EVOLUTION IN ACTION

* Birth, sex, and death are all experiences that involve individual living things, even if it sometimes takes two to make them happen. Evolution, on the other hand, is not like this. Instead of affecting individuals—which eventually die—it acts on things that persist across the generations.

sex is just a way of copying genes

MEAN GENES

* The ultimate units of evolution are individual genes. Taken to its extreme, this means that living things are simply vehicles that genes use to copy themselves and spread—a view that was originally advocated by the British zoologist **Richard Dawkins** in his book *The Selfish Gene.* However, genes do not exist independently, like candy waiting to be picked out of a bag. *As a result, the processes driving evolution work on a whole series of levels, from individual genes to entire species.*

no one's having any of my candy

selfish gene

STAYING BALANCED

In 1908, a British mathematician and a German doctor independently came up with mathematical proof showing that sexual reproduction cannot—on its own—cause evolutionary change. Known as the **Hardy-Weinberg Law,** it shows that the frequency of different alleles in a large closed population remains the same over successive generations, even though the alleles are recombined when they are handed on. *The "Hardy-Weinberg equilibrium" means that evolution is not a self-generating process.*

DIVIDED WE STAND

✱ One of the most important of these levels is a population. Populations exist because the members of a species cannot all be in the same place at the same time and therefore cannot fully interbreed. Instead, they form groups that are partially or fully separate. In some species, a population can consist of just a few individuals. In others—such as locusts—single populations can be millions strong.

✱ This separation has important implications. *If the members of a population breed with each other on a free-for-all basis but rarely breed with outsiders, their population develops an isolated pool of genes. Over a long period, isolated gene pools are subject to local influences, so tend to change in distinctive ways, and the population evolves.*

interacting with other forms of life

NEIGHBORLY RELATIONS

✱ *On a higher level, selection often operates by shaping the way organisms interact with other forms of life.* Lions eat zebras; bees pollinate plants; and microscopic algae live inside corals where they provide food in return for protection. Interactions like these involve both separate populations and separate species, amplifying the scope for selection and change.

KEY WORDS

POPULATION:
a group of individuals belonging to one species that occupy the same area at the same time

GENE POOL:
the complete collection of alleles that are present in a population

bees pollinating

SURVIVING THE STORM

I can't even see the road

which sparrows will survive the storm?

* In 1889 the American biologist H. C. Bumpus studied the effects of a storm on a population of local sparrows. He showed that the sparrows that survived tended to be those with middle-of-the-road characteristics—one of the first documented examples of "stabilizing selection."

MR. AND MRS. AVERAGE

'average'

* *Natural selection can act in one of three ways.* As Bumpus discovered, *it often favors the average by penalizing individuals that depart from the optimum range*. Bumpus found that long-winged and short-winged sparrows were disproportionately affected by the storm, while birds with average-length wings fared best.

* Since Bumpus' research was published, this form of natural selection has been found to apply in a wide range of living things. *However, selection does not always act in this way. In some cases it can favor individuals that lie at one extreme, or it may favor several extremes at once.*

More and more...

Directional selection is a powerful force for change in the natural world. Its manmade counterpart—artificial selection (see page 61)—means plant and animal breeders can exaggerate particular characteristics in any desired direction. In a long-running experiment begun in 1896, plant breeders at the University of Illinois have repeatedly selected corn for a high oil content to see how far directional selection can continue. A century later, after more than 80 generations, the oil content has more than quadrupled and is still increasing.

OUT ON A LIMB

✱ A consistent drive toward one extreme is known as <u>DIRECTIONAL SELECTION</u>. *When this kind of selection applies, the characteristics of a population steadily shift away from the original average. Unlike stabilizing selection, which can theoretically continue indefinitely, directional selection eventually reaches a stage where no further advantage is conferred by being more extreme. At this point, the average often settles down at the new position.*

✱ Giraffes provide a good example of how this works. At some time in the past, selection started to favor ancestral individuals with long necks, because they would have been able to reach more food. As a result, they left more offspring, and the average neck length shifted upward. This trend would have continued over a large number of generations, until the disadvantages of having a long neck—such as problems with drinking—started to make themselves felt. This would have acted as a brake on further neck development, until an optimum neck length was reached.

✱ Assuming that giraffes' necks have stopped getting longer—which may not necessarily be true—this optimum length is the average we see today.

LARGER AND LARGER...

Directional selection has been behind the increasing size of many living things. This is because large size often confers a range of adaptive advantages, such as increased effectiveness at tackling prey. However, trends like these are not the same as orthogenesis (see page 96), because they do not follow a preset path. As soon as further increase becomes a disadvantage, the increase stops.

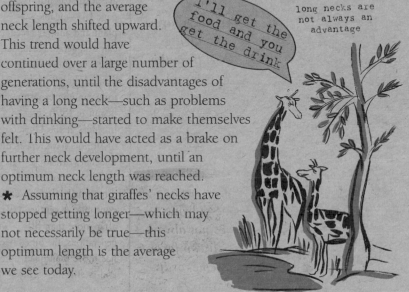

I'll get the food and you get the drink

long necks are not always an advantage

butterfly

POLYMORPHISM

* With stabilizing selection and directional selection, the variations in a population are clustered around a single average value. But in a third form of natural selection— DISRUPTIVE SELECTION—the outcome is quite different. When this force acts over many generations, it creates and maintains a limited number of distinct forms in the same species, a situation known as BALANCED POLYMORPHISM.

SPLITTING UP

* Polymorphism is a widespread feature of living things. Some snails have a variety of shell patterns, while many butterflies have a number of different wing patterns. Humans are also polymorphic. Our polymorphisms include two major blood groupings—the ABO and Rhesus systems—and nearly 300 minor ones.

* Polymorphism is most apparent in major differences like these, but it also includes a whole host of lesser variations,

snow goose

snail

produced by the maintenance of a range of alleles for different genes. *Whatever its scale, a feature of polymorphism is that the separate variants, or* MORPHS, *do not intergrade. There are no halfway houses between two morphs.*

THE GREAT DIVIDE

* As organisms adapt to their environment, they might be expected to become genetically more alike. But with species that show balanced polymorphism, this clearly does not apply. So what is it that maintains different variants in an interbreeding population?

* This question is one that biologists have yet to fully answer. It may be useful to be in a minority—rare morphs, for example, are less likely to be recognized as prey. A more complex possibility is that polymorphism is an inevitable side-effect of HETEROZYGOTE ADVANTAGE. This phenomenon (also called HYBRID VIGOR,) gives individuals with mixed alleles improved fitness in the struggle to survive. Finally, polymorphism may be maintained because selective pressures vary from one part of the population's range to another.

This is a crucial side of polymorphism, because it can form a springboard in the evolution of new species.

THE PROS AND CONS OF SEX

The existence of different sexes is the most striking example of polymorphism in living things. In most animals, sex is determined by the presence (or sometimes absence) of special "sex chromosomes." In humans, individuals with two X chromosomes are female, while ones with an X chromosome and a Y chromosome are male. *In evolutionary terms, sex is a costly development, because in most cases only one sex is able to produce young. Offsetting this is a major advantage: the genetic reshuffling that occurs during sexual reproduction produces new and potentially useful combinations of genes.*

butterfly

GENES ON THE MOVE

* A cluster of trees growing on a remote island may seem like a commonplace example of a perfectly isolated population but, in nature, truly isolated populations are extremely rare. With trees on an island, a single pollen grain drifting in on the wind can ship in new genes that may encourage evolutionary change.

a perfectly isolated population?

Stranded

If Apollo butterflies do not cross low ground, how did they become split into separate populations in the first place? The answer is that Apollos are adapted to subalpine conditions, which were common across much of lowland Europe at the end of the last ice age. As the climate warmed, the butterflies were forced to retreat uphill, creating the fragmented distribution that exists today.

an expert can identify characteristic wing patterns like a fingerprint

SWAP SHOP

* The transfer of genes between isolated populations is called gene flow. In some species, particularly ones that reproduce with the help of windblown pollen, gene flow is very pronounced. As a result, there is always a good chance of new alleles being introduced from outside.

* In other species, the rates of gene flow are much lower, because individuals are not much good at making contact with anything outside their normal range. For example, Apollo butterflies live in mountains throughout Europe, forming more than a dozen scattered populations.

These populations are separated by low-lying ground, which Apollos seem reluctant to cross. As a result, each population of Apollos has kept its gene pool largely intact and has evolved characteristic wing patterns that an expert can identify like a fingerprint.

LUCK OF THE DRAW

* Another factor that influences populations is GENETIC DRIFT. This is a random process, and it takes place because individuals vary, which means that alleles are not evenly spread.

* *When the members of a large population reproduce, the overall gene pool remains unchanged. But if the population is very small (say 50 individuals or less), chance effects begin to creep in, and it is always possible that individuals with rare alleles will happen to be the ones that reproduce. As a result, their characteristics will become more common— not through natural selection, but purely by accident.*

* Genetic drift is difficult to measure in nature, because it is often masked by other factors. However, in small populations it can have a major effect.

whoops, one extra!

THE FOUNDER EFFECT

Genetic drift plays an important part in THE FOUNDER EFFECT, which comes into play when a small number of parents manage to set up home in a new area, such as an island. When this happens, the gene pool of these "founders" is unlikely to have exactly the same balance as the gene pool of the species as a whole. The Amish farmers of Lancaster, Pennsylvania, show a striking example of the founder effect. Founded by three couples in the late 1700s, its members have a high rate of polydactylism (being born with extra fingers). In this group, the incidence of polydactylism is millions of times higher than in the world's population as a whole.

STRICTLY NEUTRAL

✱ At one time, geneticists could only investigate polymorphism by looking at its visible effects. But during the 1960s new technology made it possible to examine the structure of individual proteins and, eventually, of individual genes. The results came as something of a shock.

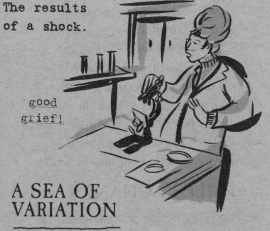

good grief!

CHARTING THE PAST

The fact that genes evolve means that they diverge as time goes by. In recent years, this has been used to map out how and when different forms of life have diverged (see page 148).

Whatever happened to natural selection?

Neutralists play down the importance of natural selection in molecular evolution. However, if selection has only a minor effect on this level—as now seems likely—this does not mean that it is merely has a walk-on part in evolution as a whole. On the level of organisms and populations, selection is thought to be the dominant force steering change in different species.

A SEA OF VARIATION

✱ *According to "traditional" neo-Darwinism, natural selection weeds out multitudes of mutations that are harmful but preserves the useful few.* Consequently, genes—and the proteins they produce—should evolve at a slow but variable rate. *However, the real picture is quite different. In most species, a large proportion of proteins are highly polymorphic, representing a wealth of variation that is difficult to explain.* The level of polymorphism is so great that natural selection alone would have difficulty keeping it in line. So how does this variation come about, and how does it manage to persist?

THE GREAT ESCAPE

★ An answer was proposed in 1967 by the Japanese geneticist **Motoo Kimura**. *Instead of labeling mutations as either good or bad, Kimura suggested that most are actually neither. Instead, they act like pieces of harmless baggage that genomes collect as time goes by. According to this idea—called the neutral theory of molecular evolution— these harmless genes spread throughout a population by random genetic drift. As they do not alter the way proteins work, they escape the effects of natural selection.*

HARMLESS CHANGE

★ When it was first proposed, the <u>NEUTRAL THEORY</u> was widely criticized. To many biologists, the idea of neutral mutations was about as plausible as the idea of neutral car parts—altering these at random would definitely have non-neutral effects. But evidence has accumulated that supports Kimura's approach. *For example, the traditional Darwinist view holds that the most important parts of proteins—the ones essential to the way they work—should be under the greatest selection pressure, so should change most often. The neutralist position is exactly the opposite, and it seems to hold true.*

<u>HOW PROTEINS VARY</u>

Proteins are made of sequences of units called amino acids—in essence, they are like strings of 20 different kinds of beads. Variation in protein molecules occurs when the usual pattern changes, and one kind of bead is substituted for another. This variation is triggered by mutations in genes, because genes control which beads are used in each string, and the order in which they are strung.

this is not a real picture of natural selection

what happens in natural selection?

HOW NEW SPECIES EVOLVE

* Despite calling his book *The Origin of Species*, Darwin said relatively little about how new species arise. When he did, he concentrated largely on what is now called VERTICAL EVOLUTION. Since his time, the way new species evolve has become better understood— but, even so, it remains one of the most widely debated issues in evolutionary theory.

Darwin said relatively little about how new species arise

ISOLATING MECHANISMS

Apart from geographical separation, a variety of other factors can prevent separate populations from interbreeding successfully. Differences in breeding seasons and courtship behavior can prevent interbreeding from occurring in the first place, while other mechanisms—such as hybrid sterility—may prevent interbreeding having a successful outcome.

THE ORIGIN OF DIVERSITY

* New species can evolve in two different ways. In vertical evolution, or ANAGENESIS, a species gradually becomes so different from its original form that eventually a new species is created. This process may be repeated many times, over millions of years—so that the species passes through a variety of different forms, each one supplanting its predecessor. *An important feature of this kind of*

courtship behavior

evolution is that it creates change without creating extra diversity: no matter how much time goes by, only one species exists.

✱ The second and more common form of speciation is called CLADOGENESIS. It takes place when an original species begins to diverge into a number of different genetic lines. Ultimately, these lines develop into new species—and once this has happened, the original species, by definition, no longer exists. *Unlike anagenesis, cladogenesis increases the diversity of life.*

SEPARATE PATHS

✱ *For an original species to split, separate lines have to become isolated in a way that prevents them from interbreeding.* The most obvious isolating mechanisms are physical barriers, such as seas and mountain ranges, that prevent any genetic flow between different populations of a species. This leads to ALLOPATRIC SPECIATION—the creation of new species through geographical separation.

✱ Some geneticists believe that speciation can also occur without any geographical separation being involved. In this situation, called SYMPATRIC SPECIATION, different populations occupy the same area, but develop differently and gradually cease to interbreed. If this reproductive isolation is maintained long enough, it eventually brings about several new species within the original species' range.

KEY WORDS

SPECIATION:
the multiplication of species through the gradual evolution of new characteristics

ANAGENESIS:
evolution within a single lineage, with an ancestral species being replaced by a succession of descendants

CLADOGENESIS:
evolution of several species from an original ancestor, creating a number of lineages

DEVELOPING DIFFERENCES
Sympatric speciation is a controversial subject. Experiments with fruit-flies suggest that mating preferences can trigger disruptive selection (see page 124) in sympatric populations; but, so far, the evidence for sympatric speciation is far from watertight.

LEAPING AHEAD

* The fossil record is like a logbook of evolution. In some places new species seem to spring up with almost clockwork regularity, while in other places bursts of speciation seem to be interspersed by long periods during which very little happens. In recent years, examination of this kind of evidence has provoked some conflicting interpretations of the pace of change.

evolution proceeds in a series of jumps

EVOLUTION BY JUMPS

* Most biologists still hold with Darwin's "gradualist" view that new species evolve in a series of imperceptible steps. *According to this idea, the major changes that occur during speciation are the result of minor changes that build up over long periods.* Putting it in more formal terms, MACROEVOLUTION is the accumulated result of MICROEVOLUTION.

* In the early 1970s two American paleontologists—**Niles Eldredge** and **Stephen Jay Gould**—challenged this belief. *Drawing on known fossil sequences, they proposed that evolution proceeds in a series of jumps, rather than* *by steady change.* According to this proposition, known as <u>PUNCTUATED EQUILIBRIUM</u>, macroevolution involves processes that stand apart from the routine adjustments produced by natural selection.

polyploid plant

EVIDENCE FROM AFRICA

✻ In 1981 a study carried out in Africa seemed to support the punctuationist view, when it showed sudden jumps in the evolution of freshwater mollusks. The fossil record showed at least two points where a whole range of species were suddenly replaced by new forms.

✻ At the time, this discovery was widely described as "disproving Darwinism," but a fuller analysis of the evidence has shown that things are not as clear-cut as they seemed. *Stabilizing selection (see page 122) means that there is nothing intrinsically un-Darwinian in long periods of equilibrium, even when it is certain that no new forms have appeared. Conversely, "sudden" jumps in the fossil record are not necessarily un-Darwinian either. Most of them take many thousands of years—ample time for small changes to accumulate.*

✻ There is still disagreement about the possible existence of evolutionary jumps—but to most biologists, macroevolution and microevolution are two sides of the same coin.

a trilobite, one of the extinct fossil arthropods

INSTANT SPECIES

In certain circumstances, new species can appear almost instantaneously. This happens when an individual's chromosomes spontaneously multiply, giving it double, triple, or sometimes more than eight times the usual number. Termed <u>POLYPLOIDS</u>, organisms like these cannot reproduce with normal members of their species, but they are able to reproduce with their own kind. Polyploid species are rare in the animal world, but common among flowering plants. This is because plants often reproduce by fertilizing themselves, so a single polyploid plant can often produce fertile offspring.

WELL, I'm staying put!

Back from the dead

Coelacanths belong to an ancient line of lobe-finned fish that were thought to have died out more than 60 million years ago. The discovery of a living coelacanth in the 1930s was one of the biological events of the century—a fishy equivalent of finding a living dinosaur.

coelacanth

RUNNING IN PLACE

✱ In Lewis Carroll's *Through the Looking Glass*, the Red Queen tells Alice that "it takes all the running you can do to keep in the same place." Many biologists now believe that this is exactly what happens in the natural world, as living things force each other to change.

TIT FOR TAT

✱ The idea is known as the RED QUEEN HYPOTHESIS, and the thinking behind it runs like this. *Imagine that two species share the same environment. One develops a new adaptation that puts the other at a slight disadvantage. This disadvantage increases the selection pressure on the other species to change as well. When it does, the cycle continues. As a result, change keeps triggering change, even though there may be no external factors driving it on.*

✱ In practice, living things usually interact with a large number of other organisms, so adaptation and counter-adaptation are much more complex than this. But if the Red Queen hypothesis is correct, it means that change is a built-in characteristic of ecosystems, affecting both species and individual populations.

starfish

IS IT TRUE?

✱ Charles Darwin clearly recognized the existence of interactive adaptation, but the Red Queen hypothesis was first put forward in the 1970s. If it is correct, it means that species may evolve and become extinct even in the absence of any environmental change.

I thought you were extinct!

a living coelacanth was caught in the 1930s

✱ An alternative view, called the STATIONARY MODEL, holds that interactive change is only a minor factor in evolution. According to this theory, major evolutionary events are driven largely by environmental change. In the Stationary world, innovation and extinction are likely to be clustered around periods when the environment is undergoing rapid change.

✱ So far, evidence has been found to support both ideas. Fossilized plankton, for example, seem to suggest that change can occur even in very stable conditions, while other parts of the fossil record show evolutionary bursts at times of change.

LIVING FOSSILS

Some complex forms of life have existed, apparently unchanged, for millions of years. They include starfish, horseshoe crabs, and also the COELACANTH— a fish that was presumed to have been long extinct, until one was hauled out of the ocean off South Africa in 1938. So has the evolutionary clock stopped for species like this? And if so, why? *The answer is that these species are almost certainly continuing to evolve— today's coelacanths, for example, are not genetically identical to those that existed millions of years ago, even if they look just like them.* However, compared with land animals, these marine species do live in highly stable environments, where there is little physical pressure for change. As a result, outward change has been very slow.

WHY BE NICE?

* In evolutionary terms, it makes good sense for a parent to look after its young because, by doing so, it increases the chances that its genes will be handed on. But, in some cases, animals help others in a way that seems like selfless generosity. In a world shaped by natural selection, how could this kind of behavior have evolved?

I'm just looking after our genes

Bad news for cubs

Kin selection has a less attractive flipside that is demonstrated by lions. If an unrelated male takes over a lion pride—as frequently happens—he is likely to kill all the pride's existing cubs. By disposing of the youngsters, he gets rid of individuals that do not share any of his genes and also encourages the females to breed again. This time, the new cubs share half his genes.

THE GREATER GOOD

* An extreme example of animal altruism occurs in honeybees. Worker bees cannot reproduce, and they die if they use their stingers. On the face of it, a worker bee's death defending its nest seems like pure heroism—a supreme sacrifice that reaps no genetic reward.

* The idea of selfish genes shows that the "sacrifice" is not quite as heroic as it appears. ***Because of the unusual way social insects reproduce—with only one female laying eggs—worker bees share an above-average number of alleles.*** In humans, for example, a child has a 50 percent chance of sharing a particular allele with one of its siblings, but in worker honeybees the figure is 75 percent.

This means that a worker bee has a better chance of propagating its genes by helping the queen to produce more workers than it would if it reproduced itself.

A HELPING HAND

kookaburra

* Elsewhere in the animal world, altruism is tied up with RELATEDNESS. The closer the relationship between two individuals, the more they have to gain by helping each other out. This is known as KIN SELECTION, and it explains why many young birds—such as kookaburras—often help their parents to feed the next brood of young.

* Sometimes, however, animals seem to help individuals that are not close relatives. Baboons, for example, help other baboons in fights, and dolphins help injured animals to the surface. The most likely explanation of such behavior is that it has evolved through reciprocation. Having gone to the aid of a non-relative, the helper may be paid back.

KEY WORDS

KIN SELECTION: a form of natural selection that favors groups of related individuals

SIZING UP YOUR RELATIVES

The number of genes that you share with your relatives can be expressed by an index of relatedness. If you have an identical twin, you share 100 percent of your twin's genes, giving an index of 1. You share 50 percent of a sibling's genes, giving an index of 0.5, and the same goes for either parent. For grandparents, uncles and aunts, and nephews and nieces, the figure drops to 0.25. For a second cousin the figure is about 0.03 and for a third cousin less than 0.01—so low that, as far as gene selection is concerned, the relationship is probably irrelevant.

dolphin lifeguard

HELP HELP

there's a different type of fitness in evolution

KEEPING FIT

✳ The phrase SURVIVAL OF THE FITTEST conjures up images of living things pushing each other aside as they scramble their way to the top. But, in evolutionary terms, fitness is not the same as being physically robust, and strength does not necessarily guarantee success.

A circular argument?

One of the criticisms often leveled at the idea of natural selection is that it is based on a circular argument: natural selection is often defined as the "survival of the fittest," but the fittest are defined as "those who survive." *Darwinists respond to this by pointing out that fitness is not so much a measure of survival ability as of success in handing on genes to future generations.* If an individual is good at surviving but poor at reproducing, its overall fitness will be poor, no matter how long it lives.

OUT IN FRONT

✳ *Fitness is a measure of how well an organism deals with the challenges of everyday life, and how effectively it manages to hand on its genes.* For an animal like a hare, being able to run fast is an obvious aspect of fitness—because if a hare is easy to catch, its life is likely to be brief. But for hares, fitness also includes a range of other characteristics, which have nothing to do with being athletic. Among these are a high reproductive rate, good hearing and eyesight, and efficient use of food.

THE MOVING TARGET

✳ Unlike the sort of fitness that develops by going to a gym, EVOLUTIONARY FITNESS it is not some fixed goal that individuals can achieve. One reason for this is that fitness has no top line. Another is that environmental changes often mean that the goalposts are on the move and, as a result, fitness shifts as well. For example, wooly mammoths had a high degree of fitness when the northern hemisphere was in the grip of the last Ice Age. However, when the climate warmed up, their fitness declined, even though they stayed the same.

there's normally a tortoise chasing me

hare

THE FITNESS CLUB

* Fitness usually applies to individuals and the genes that they possess. *However, in animals that habitually help their close relatives (see page 137), fitness is something that can also encompass a group.* In these groups, an individual's fitness may not be that impressive: an ant on its own, for example, does not present much of a threat to its enemies and may also have trouble finding food. But, as a group, an ant colony is a formidable fighting force and a highly effective food-gathering machine. Ants and other social insects have become some of the most successful animals on Earth; with the possible exception of humans, their INCLUSIVE FITNESS is second to none.

I'M SO UNFIT

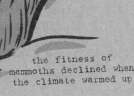

the fitness of mammoths declined when the climate warmed up

139

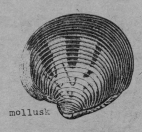
mollusk

SHELL SHOCK

Convergence sometimes produces such striking similarities that underlying relationships become difficult to establish. Barnacles—which were studied intensively by Darwin (see page 67)—look very much like limpets, because they are protected by hard plates that make up a conical shell. However, closer examination shows that a barnacle has tiny jointed "legs," a feature that limpets and other mollusks have never evolved. Barnacles are in fact aberrant crustaceans—a fact that eluded naturalists until the 19th century.

SHARED SOLUTIONS

* Evolution tends to make living things increasingly different and increasingly specialized for particular ways of life—a fact that Charles Darwin recognized when he wrote *The Origin of Species*. But within this overall pattern of divergence, natural selection sometimes seems to go into reverse, by making different species more alike.

where would I be without evolution?

EVOLVING UNDERGROUND

* *This phenomenon is known as* <u>CONVERGENCE</u>, *and it occurs when natural selection acts on unrelated organisms that share similar ways of life*. For example, moles spend most of their lives underground and have large spadelike front paws which they use to dig their tunnels. Their thick finger bones ensure that they can shift large amounts of earth quickly, while their cylindrical bodies help them to move through their tunnels.

* In the insect world, mole crickets share a similar way of life. Although their front limbs are constructed in a completely different way to a mole's, they, too, have spadelike "feet" for shoveling earth, and their bodies are also cylindrical. *Through convergence, two very different animals have evolved similar adaptations for getting around underground.*

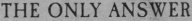

cacti are found only in the Americas

THE ONLY ANSWER

* Spadelike feet are not the only evolutionary solution to the problems of tunneling—mole rats, for example, dig with their teeth. *But, in some situations, selective pressure toward one particular solution is very strong, and convergence is much more marked. The classic example of this involves terrestrial vertebrates that have taken up life in water.*

* Water is almost a thousand times denser than air, and is much more viscous. For animals originally adapted to life on land, moving around in water is hard work, because friction soaks up large amounts of wasted energy. The answer to this problem is to be streamlined, and to move with the help of flattened fins. Thanks to convergence, today's most fully adapted "returnees"—whales and dolphins—have a completely fishlike outline.

Cactus lookalikes

Convergence accounts for the similarity between cacti and a range of unrelated desert plants. With the exception of a single species, cacti are found only in the Americas. The deserts of Africa and Asia have some cactuslike plants, with the same barrel-shaped stems for storing water. However, none of them are members of the cactus family.

mole crickets are adapted for getting around underground

parallel evolution

PARALLEL LINES

✳ Because living things cannot throw off their evolutionary past, convergence is a largely superficial process. Unrelated species may come to look alike, and yet, deep down, the evidence of separate lines of development remains. But sometimes the same adaptations can arise independently in species that are closely related. This is known as parallel evolution.

KEY WORDS

SECONDARY CHARACTERISTIC: one that has appeared relatively recently in the evolutionary history of a species

PRIMARY CHARACTERISTIC: one that has a much longer history, and which is common to a species and its ancestors

termite

EVOLVING IN STEP

✳ *Parallel evolution is a form of convergence, but one that applies to species that share a common ancestry. It takes place when natural selection brings about a new shared feature that the ancestral species did not have.* Unlike convergent evolution, parallel evolution can be tricky to establish, because it has to be shown that the feature was not simply developed by the ancestral species and then handed on.

✳ A clear-cut example of parallel evolution can be seen in mammals that eat ants. Modern mammals divide into three groups—placentals, marsupials, and monotremes—and each of these groups contains ant-eating species. They all have

giant anteater

long snouts and long sticky tongues, and they all have strong claws for digging ants' nests open.

* Fossil evidence shows that none of these features was present in the ancestral mammals that eventually gave rise to the mammal groups existing today. As a result, the ant-eating adaptations must have evolved independently—a case of related species adapting in parallel.

Parallel lives

The social lifestyles of some insects are a result of parallel evolution. True social life, which involves division of labor, has evolved in two separate groups of insect: hymenopterans (which include ants, bees, and wasps) and termites. Instead of dispersing when they reach adulthood, social insects form permanent family groups; in most cases only one individual, the queen, lays eggs. The parallels between ants and termites are particularly strong—which explains why termites are sometimes called "white ants."

"I think I'll have the ants"

TANGLED TRUNKS

* Parallel evolution can produce complex and confusing results. Trees, for example, all have a tall woody stem that allows them to reach higher in the struggle for light. Common sense might suggest that trees evolved once, and then diversified.

* *In fact, trees have evolved many times*. Of the 250 or so families of flowering plants, there are some that contain only trees, some that contain a mixture of trees and non-woody plants, and some that contain no trees at all. Among flowering plants, the treelike form is a SECONDARY CHARACTERISTIC—one which developed at a relatively late date.

queen

drone

EVOLVING TOGETHER

* In the natural world, very
few things live on their own.
Instead, most are surrounded by
other forms of life, which make
up part of their environment.
If two species frequently
interact, their evolutionary
paths may begin to intertwine,
with each being influenced by
the other's adaptations.

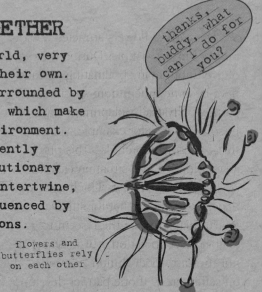

thanks,
buddy, what
can I do for
you?

flowers and
butterflies rely
on each other

in the natural
world, very few
things live on
their own

THE LURE OF FLOWERS

* *This interlinked development is called*
<u>COEVOLUTION</u>. *It has occurred throughout
the whole history of life, but some of its
most dramatic effects have been felt in the
last 100 million years, with the explosive
growth in partnerships between animals
and flowering plants.*

* Flowers are in effect living
advertisements, designed to attract
animals. A flower's animal visitors—
usually insects, but sometimes birds or
bats—distribute the flower's pollen and,
in return, receive a small but useful meal,
usually in the form of nectar. This
arrangement allows plants to overcome
the limitations of being rooted in one spot
and provides animals with a reliable
source of food.

✱ The earliest flowers attracted a wide range of insect visitors. During the course of evolution, many pollinating partnerships have become much more specific, with partners physically adapting to each other. In some cases—for example, yuccas, fig trees, and many orchids—this process of adaptation led to a situation called OBLIGATE MUTUALISM, in which each species depends on its partner for survival.

✱ Partnerships like these seem to involve deliberate cooperation, but in reality that is not the case. Coevolution is founded solely on self-interest. If one partner develops an exploitative edge, natural selection sees to it that the partnership becomes more and more one-sided.

✱ *Nature is full of examples of coevolved relationships that are more like arms races than cozy collaborations.*

that seals the partnership then

For instance, plants have evolved batteries of chemicals to fend off leaf-eating animals, while animals have, in turn, evolved new ways of breaking down these potentially harmful substances.

COME TOGETHER

The word SYMBIOSIS is used to describe any close association between members of two different species. In a MUTUALISTIC ASSOCIATION both partners benefit; in a COMMENSAL ASSOCIATION one partner benefits, but the other is neither helped nor harmed. In a PARASITIC PARTNERSHIP the benefits are all one-way; the parasite lives at the expense of the host.

An OBLIGATE PARTNERSHIP is one that is essential for an organism's survival. A FACULTATIVE PARTNERSHIP is one that it may join, but which it can manage without if it had to.

In SYMBIOTIC PARTNERSHIPS there is often scope for cheating. Some pollinating insects and birds save time and energy by biting through the backs of flowers to steal nectar. They exploit the food that the flowers produce but provide nothing in return.

LUCKY BREAKS

✱ Life, it is often said, isn't fair. Evolution is even less so, and in many cases the difference between success and failure ultimately comes down to luck. In evolutionary terms, good luck often comes in the guise of environmental changes that favor the fortunate few.

YES I'VE WON!!

evolution is often a matter of luck

ACCIDENT VERSUS INEVITABILITY

There is debate among biologists about the importance of CONTINGENCY— chance events and favorable coincidences—in evolution. *If contingency plays a major role in evolution, life as it exists today is largely the result of historical accidents. If, on the other hand, it plays only a minor role, this suggests that some features of life may be an inevitable outcome of evolution.*

READY AND WAITING

✱ One form of biological lucky break involves a phenomenon called PREADAPTATION. *It occurs when a characteristic that originally had some use under one set of conditions turns out to be useful in a different way when conditions change. Despite its name, preadaptation does not imply advance preparation, because natural selection cannot anticipate future events. It simply redeploys existing adaptations, sometimes to great effect.*

✱ One example of this can be seen every spring in northern Europe, when millions of house martins migrate north to breed. House martins originally nested on cliffs—but when, a few thousand years ago, houses started to appear in Europe, martins began to use the eaves of roofs as nesting sites. This behavioral preadaptation has proved to be very advantageous, because house martins can now nest in places where there are no cliffs at all.

house martins originally nested on cliffs

CLEARING THE STAGE

* *Lucky breaks can also be triggered by climate change, by changes in sea level, and by the gradual shifting of the continents. They can also be brought about by unusual weather patterns, which can help animals to spread to regions that were originally beyond their reach.* Strong tailwinds may have helped the American monarch butterfly to gain a toehold in Europe, and were probably involved when the ancestors of Galapagos finches (see page 56) wandered out into the Pacific.

* *However, the ultimate lucky break comes in surviving catastrophes that wipe out competing species.* The dinosaurs almost certainly died out when a meteorite hit the Earth, but pocket-sized mammals managed to survive—*a stroke of good fortune that allowed them to take over the empty stage and ultimately to dominate life on land.*

Surviving DDT

The variability of living things means that there is a vast potential for preadaptation. In insects, for example, a number of genes happen to confer resistance to the pesticide DDT. When DDT began to be used in the 1940s, these genes would have given their owners an immediate and powerful selective advantage. As a result of preadaptation and selection, DDT-resistant insects are now much more widespread than before.

some insects were immune to DDT

CHAPTER 5

CLASSIFYING LIVING THINGS

* Despite its less-than-glamorous reputation, <u>CLASSIFICATION</u> is a key part of modern biology. It enables researchers to identify living things in a precise way, and to show how they are thought to be related by evolution.

arranging
organisms in
a series of groups

LUMPERS AND SPLITTERS

<u>TAXONOMY</u> and train-spotting have one feature in common: the more you look, the more you find. As a result, experts in particular areas of biology often notice subtle differences in living things that others would overlook, leading them to split groups into large numbers of species. On the other side of the fence to these "<u>SPLITTERS</u>" are the "<u>LUMPERS</u>"– biologists who believe that for many species, distinctions are imaginary rather than real.

LASTING WORK

* The basis of modern classification was laid down by Linnaeus (see page 34), who devised a system of binomials, or two-part scientific names. These names are based on Latin. Although they can look daunting, they have two important advantages: they identify organisms with much greater precision than common names and they are completely international, staying the same whatever language you happen to speak.

* *In addition to identifying individual organisms, Linnaeus also arranged them in a series of groups, based on similarities that they show. Although he did not believe in evolution, his twin-track approach— naming and grouping—turned out to be almost purpose-made for tracing evolutionary relationships.*

THE LIBRARY OF LIFE

✱ Modern classification works like an all-encompassing filing system, with a hierarchy of folders (the taxa) that nest inside each other. The fundamental TAXON—and therefore the smallest type of folder—is the species. Species are enclosed in a succession of larger folders, ranging from GENERA to KINGDOMS. Kingdoms are traditionally the largest divisions of the living world.

✱ At the species level, the name that goes on the folder is more than just a label. It also functions as an address, showing which higher taxa a species belongs to, and therefore which other species are its closest relatives.

✱ An ideal classification system should reflect evolution as closely as possible, with related species being grouped together, and more distant ones farther apart. Unfortunately, this ideal is not always easy to achieve. **Deciding where a species belongs involves making judgments about its origins, and this makes classification a partly subjective process.** In recent years, new methods have removed some of this subjectivity, but classification schemes are updated all the time, as new information comes to light about the paths evolution has followed.

TAXON (PLURAL TAXA): a category used in biological classification (the principal taxa are species, genus, family, order, class, phylum, and kingdom)

ah yes mops mops will do admirably

choosing a name

Choosing a name

Custom dictates that the discoverer of a new species chooses its scientific name. Names often describe a salient feature of the species or indicate its distribution, but some are rather more obscure. One species of bat, for example, is named *Syconycteris hobbit*, and another is called *Mops mops*.

TRACING THE PAST

* By tracing the path that evolution has followed, biologists can throw some light on how living things came to be as they are today. However, thanks to missing records and confusing clues, the process is not nearly as easy as it sounds.

AMBIGUOUS EVIDENCE

* Two kinds of evidence help in piecing together the path of evolution. The first is built into the present structure of living things, while the second consists of fossils. If they are correctly interpreted, similarities and differences can give some idea of how closely living things are related, while fossils reveal forms of life that existed in the past. Unfortunately, as evidence goes, both can be hard to read. This explains why assembling PHYLOGENIES—or evolutionary pathways—is a tricky and sometimes controversial business.

UNCOMFORTABLE GROUPS

* The living evidence sometimes seems straightforward. For example, birds share a clutch of features that other animals do not have, so it makes sense that they must be

WHO'S AHEAD?

Biologists often loosely describe species as being PRIMITIVE or ADVANCED, suggesting that some have made a better job of evolving than others. Although convenient, these words can be misleading, because in evolutionary terms a primitive species may be just as effective as an advanced one in meeting the challenges of everyday life. *"Primitive" is a shorthand way of saying that a species has changed relatively little in its evolutionary history, while "advanced" means that a species has accumulated a wider range of new features.*

the piecing together of evolution

related to each other. In other words, they make up a <u>MONOPHYLETIC GROUP</u>— *a collection of living things that have originally sprung from the same distant ancestor.* If a complete phylogeny of birds could be constructed, it would have that shared ancestor as its starting point.

"so we are related then"

✱ But what about fish? Superficially, they also seem to share a specific set of features, but closer inspection shows that this is not so. Sharks and rays are quite different from bony fish, and jawless fish are even more distinct. This anatomical evidence shows that fish are a <u>POLYPHYLETIC GROUP</u>—*one containing species that have evolved from quite separate starting points.*

FOSSIL FACTS

✱ Fossils can help to sort out polyphyletic groups, but they can also create problems of their own. *Unlike living things, most fossils only record hard structures, so vital evidence is often lost.* The fossil record is full of gaps and has no built-in signposts, making it difficult to tell whether a particular fossil is an ancestor of species existing today or has left no living descendants. In many cases, the fossil trail soon runs cold.

it's an evolutionary tale

KEY WORDS

PHYLOGENY:
the evolutionary history of a group of organisms, often shown as a "family tree"

MONOPHYLETIC:
a group of species that have evolved from a single ancestral species

POLYPHYLETIC:
a group of species that have evolved from more than one ancestral species

---

151

CLADES AND GRADES

* During the 1960s two new methods were developed for classifying living things and reconstructing the path of evolution. In different ways, each tries to make classification more scientific and less prone to differences of opinion.

DOING IT BY NUMBERS

* One method, called <u>NUMERICAL PHENETICS</u>, does away with subjective decisions altogether and turns classification into an exercise in mathematics. In essence, the procedure is simple. Take, for example, a dog, a cat, and a chicken. Give all their characteristics equal weight, then add them up and compare them. If the dog and cat share 4,990 characteristics, but the dog and chicken only 3,586, then the dog is more closely related to the cat.

* The system groups organisms by the features they have at present, rather than by their past, and it does not make any allowances for convergence (see page 140). However, simply by processing large numbers of features, it can give an idea of how species are related.

3,586 characteristics

4,990 characteristics

PARING DOWN

Comparisons between many species can produce a large number of possible cladograms. When this is the case, the PRINCIPLE OF PARSIMONY is applied. This holds that the cladogram most likely to reflect the actual path of evolution is the one that features the fewest branches.

FORKING OFF

✻ The second method, called <u>CLADISTICS</u>, was introduced by the German entomologist **Willi Hennig**. It focuses on evolutionary innovations and uses them to construct branching trees called <u>CLADOGRAMS</u>. Unlike a typical "family tree," the branches in cladograms are always dichotomous, and each one marks a single evolutionary innovation. Species that share the highest number of <u>APOMORPHIC</u> (recently derived) innovations are separated by the lowest number of branches, showing that they are the most closely related.

✻ *In cladistics, the only groups to be given formal recognition are* <u>CLADES</u>*—ones that include an ancestral species and all its descendants.* Birds, for example, make up a clade, and so do penguins or owls. However, flightless birds do not, because they have evolved on a number of separate branches in the bird clade. Instead, they represent a **structural grade—a similar state of development reached by related but different evolutionary lines.**

✻ When cladistic analysis was first introduced, its rigorous approach provoked ridicule and even outrage. Today, most of the indignation has died down, and the concept of clades has become an indispensable part of classification.

KEY WORDS

PHENETIC:
concerned with features that can be observed and measured

APOMORPHIC FEATURE:
a feature that has evolved relatively recently

PLESIOMORPHIC FEATURE:
a long-established feature that is shared by all the members of a clade

penguins and owls belong to different clades

153

THE MOLECULAR CLOCK

* Cladistic analysis helps to show how different forms of life have diverged, but it cannot say when each split occurred. To find out, biologists can now turn to a new form of evidence— the record that is written in proteins and genes.

how can we measure evolutionary time?

DNA hybridization

DNA strands from different species can be compared directly by HYBRIDIZATION. In this technique, the DNA molecules are cut up into short fragments and then treated to make their strands separate. When the two types of DNA are mixed together, any sections that have similar sequences bind together.

The strength of this binding depends on the degree of similarity, and it can be measured by heating the mixture until the fragments eventually split apart.

MORPHOLOGY VERSUS MOLECULES

* Until the discovery of PROTEIN POLYMORPHISM, dates in evolution were established indirectly, by comparing the age of fossils. When the neutral theory of molecular evolution was put forward in the 1960s, a completely new way of dating became available. Instead of dealing with changes in overall structure or morphology, this dating system looks at the structure of individual molecules.

* The neutral theory holds that harmless mutations build up in genes—and therefore in proteins—at a slow but steady rate. In the case of the protein alphaglobin, for example, this rate is about one amino-acid change per 5 million years. *If you compare* *alpha-globin molecules in two different species, add up the number of amino-acid differences and then multiply by five, the figure produced is the approximate date, in millions of years, when the two species set off on different evolutionary paths.*

WOBBLES IN TIME

✱ Some geneticists thought that the "MOLECULAR CLOCK" would do away with the need to spend hours poring over tiny differences in physical features. Chemical analysis would make evolutionary history drop into place. However, experience has shown that the clock is not without its faults. The "ticking rate" is not always easy to establish; and confusingly, genes, just like physical features, can sometimes come to resemble each other through convergence. *But despite these problems, molecular dating has become a powerful tool for studying evolutionary history.*

CHOOSE YOUR PROTEIN

ALPHAGLOBIN forms part of hemoglobin, the protein in red blood cells that carries oxygen. As a group, globins are useful in molecular dating because they are found in all vertebrates and date back at least 500 million years. A protein called CYTOCHROME C, which is used for handling energy, is even more ancient. It dates back at least 1.5 billion years and is found in practically all living things, making it an almost universal indicator for evolutionary relationships.

bullfrog

Distant relations

The human version of cytochrome c differs from a rhesus monkey's in just one amino acid out of a total of 104. We differ from whales in 10 amino acids, bullfrogs in 18, and yeasts in 45.

so it's one in every 5 million then

alpha-globin has one amino-acid change every 5 million years

HOW MANY KINGDOMS?

how many kingdoms have we now, dear?

✳ In classification, disagreement about how living things should be split up goes right to the top. Early naturalists divided life into just two kingdoms—plants and animals. Today, most taxonomists recognize five kingdoms, but some claim that there should really be a dozen or more.

WHAT ABOUT VIRUSES?

VIRUSES are not featured in classification systems, because they do not have cells, cannot use energy, and cannot reproduce on their own. Unlike cells, viruses can often be dried out and crystallized, and may be stored for years on a laboratory shelf. *If a crystallized virus is dissolved in water and then brought into contact with its host, it immediately "comes back to life" and can again infect living cells.*

FIVE-WAY SPLIT

✳ Kingdoms are defined on the basis of fundamental differences in the way living things work. The animal kingdom—which is the largest in terms of species—consists of multicellular organisms that get energy by ingesting food, while the plant kingdom consists of multicellular organisms that get energy directly from light. Fungi form a third kingdom. Unlike plants, they do not have any light-harnessing ability and instead live by absorbing simple nutrients from their surroundings.

✳ These three kingdoms account for all living things that have more than one cell.

fungi form the third kingdom

In the FIVE-KINGDOM SYSTEM, single-celled organisms make up two further groups—protists and monerans. PROTISTS have complex cells, much like those of plants or animals, while MONERANS

typhoid bacteria

are much simpler than other living things and also much smaller. Better known as bacteria, monerans are the most abundant organisms on Earth and can be found everywhere from polar ice to human skin.

SUPERKINGDOMS

* The five-kingdom system was widely adopted after it was first proposed in the 1950s, but it has since begun to show signs of strain. In the early 1980s, **Carl Woese** and his colleagues at the University of Illinois discovered that there are two fundamentally different groups of bacteria, which must have diverged long before more complex life evolved. In 1990 they proposed a system of three SUPERKINGDOMS or DOMAINS. The first two domains contain bacteria, while the third contains EUKARYOTES—forms of life that, unlike bacteria, have evolved complex cells.

* Woese's system has yet to be fully accepted, but its logic is persuasive. As more is discovered about the simplest forms of life, the highest levels of classification look certain to change.

BLURRED BOUNDARIES

Even in the five-kingdom system, some organisms do not have a single slot. Green algae, for example, straddle the boundary between protists and plants, while some fungi are single-celled and are therefore often classified as protists. *The protists themselves are a very mixed bag—in fact, they are so varied that some microbiologists believe they should be split up into several kingdoms.*

Getting real

Like other problems in classification, the kingdom question centers on a paradox. Taxonomists try to arrange organisms into "natural" groups—but, with the exception of species, taxonomic levels are purely mental constructions that do not have any concrete existence in the natural world. As a result, their scope is arbitrary and prone to change.

CHAPTER 6

HOW LIFE BEGAN

* In 1953, the American chemist Stanley Miller carried out an experiment designed to mimic conditions on the early Earth. A week later, after flashing electric sparks through a gas-filled flask, he found that he had created some of the chemical building blocks of life.

making life in a test tube

MILLER'S EARTH ON A BENCH

Stanley Miller's experimental apparatus consisted of a sealed glass system with a chamber holding a l-gallon "atmosphere" and a small water-filled flask acting as the sea. The atmosphere contained methane, ammonia, hydrogen and water vapor, but it excluded oxygen, which was not present during Earth's early history. Sparks produced by a pair of electrodes imitated lightning—one of the sources of energy that would have driven early chemical reactions.

A GAME OF CHANCE

* Miller's experiment provided the first tangible evidence that life could have arisen by chance, through a series of random chemical reactions. However, despite the widespread astonishment that greeted his results, he had only formed life's simplest precursors. For life itself to have taken shape, over 3.8 billion years ago, these building blocks must somehow have become organized structures, based on molecules that could copy themselves.

* On the face of it, the odds against this happening are insuperably vast. A single protein molecule, for example, may have 100 amino-acid units, all of which have to be present in the correct order for the protein to work. As there are 20 different kinds of amino acid, the chance of them forming the protein at random are about

1 in 10^{130}—a figure much greater than the total number of atoms on Earth. But would life have begun with complex chemicals like these? The answer is almost certainly no, which drastically shortens the odds against it originating by chance.

WHAT CAME FIRST?

✱ Today, life is based on DNA. DNA can copy itself, but it can only do this with the help of enzymes. Enzymes, however, are made using information held in DNA. *This chicken-and-egg situation is one reason why DNA may not have been the first chemical replicator.*

✱ A more likely candidate is RNA, a relatively compact nucleic acid that living things now use in protein

DNA or RNA? which came first?

synthesis. RNA molecules—or rather their simpler predecessors—would have acted as "NAKED GENES," capable of attracting and assembling the chemical bases they needed for replication. Once copying was established, these naked genes would have faced increasing competition for bases, and a form of selection would have begun. Any useful mistakes would have been preserved and copied themselves, starting the long evolutionary road to living things.

Life from outside

A small number of scientists, including the British astronomer **Fred Hoyle**, believe that life originally arose elsewhere in the Universe, arriving on Earth from outside. Backing up this theory, called PANSPERMIA, are observations showing that a variety of organic chemicals exist in space. Panspermia has attracted few converts, partly because it would be difficult for living things to survive a journey through space. A more fundamental problem is that it does not answer the question of how life arises—it simply moves it elsewhere.

"it must have been panspermia"

did life come from somewhere else?

HISS SPUTTER HISS

THE FIRST CELLS

* In a world rocked by storms and volcanic eruptions, and broiled by intense ultraviolet light, early self-copying chemicals would have faced the constant threat of disruption. Survival ultimately hinged on a crucial development: chemical membranes that shielded them from the outside world.

volcanic eruptions rocked the earth

LIVING ON LIGHT

The first living things used existing chemicals as sources of energy, but about 2.5 billion years ago, bacteria evolved that could harness energy directly from sunlight. By carrying out this process—called PHOTOSYNTHESIS— they produced oxygen as a by-product. Photosynthesis gradually transformed the atmosphere, allowing aerobic or oxygen-using cells to evolve.

GETTING UNDER COVER

* *For more than 3.5 billion years, flexible ultra-thin membranes have been a shared feature of all living things. By forming cells, membranes create miniature environments whose internal conditions differ from the environment around them.*

* Early membrane-bound structures, like the first organic chemicals, almost certainly came about through random events. Experiments have shown that they can develop where waves stir up a surface film, and they can also be produced by PROTEINOIDS—proteinlike substances formed when a mixture of amino acids is heated. *If self-replicating substances managed to become incorporated into structures like these, their chances of survival would have been dramatically improved.*

NEW CELLS ON THE BLOCK

★ Like today's bacteria, the first fully-fledged living things had very simple cells. Known collectively as <u>PROKARYOTES</u>, they monopolized life on Earth for more than 2 billion years. *But, at about this point, much more complex cells began to appear.* Unlike prokaryotes, these newcomers—called <u>EUKARYOTES</u>—were larger and much more complex. Their cells were equipped with structures called <u>ORGANELLES</u>, enclosed in membranes of their own.

★ Organelles are a eukaryotic cell's departments, and each one carries out a range of specific functions. The <u>NUCLEUS</u>, for example, stores genetic material. <u>MITOCHONDRIA</u> release energy from chemical fuel, while <u>CHLOROPLASTS</u> absorb energy from light. Although they are parts of cells, they are strangely autonomous: mitochondria and chloroplasts have their own DNA and, instead of being assembled from scratch, they multiply by splitting in two.

★ *For most biologists, these remarkable features point to a inescapable conclusion: eukaryotic cells evolved by "swallowing up" prokaryotes, forming partnerships that have lasted ever since.*

cell membranes are semi-permeable

KEY WORDS

PROKARYOTE: an organism that has simple cells without membrane-bound organelles

EUKARYOTE: an organism with complex cells that have membrane-bound organelles

Life's leaky linings

Cell membranes are not simply barriers, like the microscopic equivalent of plastic bags. Instead, they are semipermeable, allowing selected molecules in or out of the cell. This selective permeability enables cells to concentrate particular substances and make them react—an essential feature if they are to function and reproduce.

glass sponge

THE CAMBRIAN EXPLOSION

* Animals first appeared about 1 billion years ago. For a large part of this time, they were entirely soft-bodied, so they left few traces of their existence. During the Cambrian period—from about 570 to 500 million years ago—all that changed, as animal life underwent a spectacular burst of evolution.

Keeping it simple

In most animals, groups of similar cells are arranged into tissues, which carry out particular functions. Sponges do not have tissues—nor does one of the smallest animals known, a species called _TRICHOPLAX ADHAERENS_. First found in an aquarium in 1883, _Trichoplax_ is about the size and shape of the period at the end of this sentence, and moves by creeping over flat surfaces. It feeds on algae, but has never been seen in the wild.

STRANGE NEIGHBORS

* The causes of the Cambrian explosion are still a mystery. Animal fossils have been found that predate the Cambrian era, but there is nothing in earlier times to match the sudden eruption in diversity that began about 570 million years ago. _During the Cambrian era, all the animal phyla that exist today became established, together with many others that flourished briefly and then disappeared._ In the Cambrian seas, sponges, crustaceans, and mollusks lived alongside esoteric-sounding creatures such as anomalocariids, amiskwiids, and rotadiscids, which vanished without leaving any known descendants.

crustacean

CHANGING TIMES

★ Several theories have been advanced for this sudden burgeoning of animal life and the intense struggle for survival that it created. One possibility is that it was not sudden at all, and that earlier animals were simply not preserved as fossils. A second explanation is that some switch in environmental conditions provided a unique opportunity for animal life to diversify. *Many biologists believe this switch involved oxygen.*

★ As living things demonstrated for over 2 billion years, life is quite feasible without oxygen—indeed, some forms of life still thrive without it today. But, without oxygen, releasing energy is a slow process. *Oxygen is highly reactive. When it became widely available, cells that could use it were able to liberate energy much more rapidly, enabling them to live in a much more active way.*

★ This is what may have fueled the Cambrian explosion. With the help of oxygen, animal evolution would have stepped up a gear, allowing animals to become bigger and more energetic. The competition for survival produced offensive weapons and hard external skeletons— creating the rich array of Cambrian fossils that survive today.

jellyfish

CLUBBING TOGETHER

Unlike simpler forms of life, animals have many cells, and these divide up the work involved in staying alive. In animals, MULTICELLULARITY probably evolved when single-celled protists grouped together to form permanent colonies, which gradually led to a division of labor. Protists called choanoflagellates have much in common with the cells of sponges, but the protist ancestors of other animals are unknown.

The garden of Ediacara

The richest collection of fossilized Precambrian animals comes from the Ediacara Hills in southern Australia. Formed about 650 million years ago, these fossils show some similarities with jellyfish. Opinions differ as to whether they are distant ancestors of jellyfish or completely different animals that eventually became extinct.

163

INVENTING THE BACKBONE

* Fewer than 3 percent of the animal species alive today have backbones— but because we are among them, backboned animals hold an oversized interest. The history of backbones shows how an apparently minor adaptation can become a major evolutionary success.

he's got no backbone that one at number 18

INSIDE OUT

Before vertebrates evolved bone, skeletons were already a widespread feature in animals, but they were all EXOSKELETONS— structures that support the body from outside. Exoskeletons work well for small animals; but, even in water, they are much too cumbersome for large animals that feed by moving around. The internal skeletons of vertebrates make up a much smaller proportion of body weight, allowing them to grow big while staying mobile.

CHORDATES

* Vertebrates—animals with backbones—belong to a phylum called the CHORDATES. In addition to vertebrates, this phylum also includes some invertebrates—baglike sea squirts and finger-sized animals called lancelets. The feature that unites all these animals is a primitive structure called the NOTOCHORD, a rod running the length of the body.

* *In vertebrates, the notochord disappears during early development. In sea squirts, it is present only in the larval stage, but its fleeting presence throws some light on how vertebrates might have evolved.*

lizard backbone

164

ACTIVE LIFE

sea squirt

✳ A sea-squirt larva looks like a tiny tadpole. The notochord stiffens its body, allowing opposing sets of muscles to flick its tail and drive it through the water. When it turns into an adult, its body changes shape and the notochord disappears. It spends the rest of its life fastened in one place.

✳ The ancestors of today's sea squirts were probably the first animals to have a notochord. Many zoologists believe that it turned out to be so useful that selection preserved it for an increasing length of time. According to this theory, the transformation from larva to adult would have been progressively delayed, until the need for the sedentary adult stage was eventually cut out altogether. *Once this had happened, the stage was set for a further major development— the evolution of bone.*

this bone is a break with the past

✳ No one knows when bone first evolved, but it marked a major break with the past. The notochord became surrounded by vertebrae, forming the central axis of a complete internal skeleton. *This kind of skeleton was far more adaptable than any that had evolved before, and it enabled vertebrates to become the largest and fastest animals that have ever lived.*

FOREVER YOUNG

The persistence of larval stages is an example of HETEROCHRONY—an evolutionary change triggered by a variation in development rates. One form of heterochrony, called PEDOMORPHOSIS, involves the development of adult reproductive features in an otherwise immature body, allowing the adult body to be sidestepped. This kind of development can be seen today in axolotls and many other salamanders. Axolotls have reached a pedomorphic halfway house, because they can breed either as swimming larvae or as air-breathing adults.

VERTEBRATE EVOLUTION

* The first vertebrates were armor-plated fish that inhabited shallow seas in Cambrian times. Reconstructions of these animals often show them in a state of open-mouthed surprise—an expression that would have been permanent, since they had no jaws.

I'd be in real trouble if I inhabited deep seas

Firstborn

Sharks were probably the first vertebrates to give birth to live young instead of laying eggs. This method of reproduction—called VIVIPARITY—gives offspring a better chance of survival, but it restricts the number that can be produced. A female cod, for example, can lay over 1 million eggs, whereas a shark rarely has more than a dozen young.

hammerhead shark

THE SUCTION FEEDERS

* *Most of these early fish worked like vacuum cleaners, sucking up smaller animals from the sea bed.* Sucking also played a part in respiration, as these animals drew water in through their mouths, then pumped it past their gills and out through openings in the sides of their heads.

* Known collectively as AGNATHANS, jawless fish declined about 350 million years ago, but they did not disappear completely. Today, they still exist in the shape of lampreys and hagfish—slimy eel-like animals that live either by sucking blood from living fish or by eating worms and dead remains.

THE FIRST BITE

✱ The first jawed fish date back to the Silurian period, which began about 440 million years ago. Though there is no fossil evidence, jaws almost certainly developed from some of the bony arches that previously supported gills. **The impact of jaws was phenomenal: instead of vacuuming up their food, fish could now bite into it.** By the end of the Devonian period, about 365 million years ago, fish included huge predators such as <u>DUNKLEOSTEUS</u>, which had platelike teeth big enough to slice a human body in two.

FROM FINS TO LIMBS

✱ Evolution was also experimenting with different ways of making fish move. Some fish—like most of those alive today—had fins that were supported by slender rays and operated by muscles stowed entirely inside the body. But in Devonian times, another fin plan evolved. Supported by bones and equipped with built-in muscles, these fins were fleshy and more maneuverable. **Fleshy fins were the forerunners of paired limbs.**

fin of a lionfish

NO BONES

The forerunners of today's sharks first appeared toward the end of the Devonian period. Unlike almost all other fish, their skeletons are made of dense rubbery cartilage instead of bone. At one time, it was thought that sharks had never had bony skeletons, but most experts now believe that they gradually lost them as they evolved.

THAT SINKING FEELING

Neutral buoyancy means that fish can stay at the same depth without having to use their fins. Sharks have large, oil-rich livers that work like buoyancy tanks—but, even so, most of them have a slight negative buoyancy, which means they slowly sink if they stop swimming. Bony fish, which make up the vast majority of species alive today, achieve true neutral buoyancy with the help of a <u>SWIM BLADDER</u>—a gas-filled "float" inside the body.

LIFE MOVES ONTO LAND

✳ The development of land-based life began more than 400 million years ago and was one of the greatest turning points in evolution. Once life had severed its dependence on watery surroundings, a completely new world was waiting to be exploited.

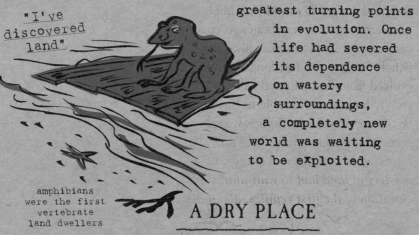

"I've discovered land"

amphibians were the first vertebrate land dwellers

A DRY PLACE

Warm-blooded life

One advantage of living in air is that it is not much good at conducting heat. As a result, birds and mammals can maintain a large temperature difference between themselves and their surroundings, allowing them to remain active even when it is cold. In water, this is much more difficult. Apart from whales and seabirds, and a few highly active fish such as tuna, aquatic animals stay at the same temperature as the water around them.

✳ For us, dry land seems an ideal environment, but it does have drawbacks. Air dries out living cells, and it is also prone to sudden changes in temperature.

✳ *The first living things to exist on land were accidental arrivals: single-celled organisms that were washed ashore, where they were poorly equipped to survive. But, when plants began to colonize muddy shores, during the Silurian period, they created the first stocks of land-based food and a strong incentive for animals to follow.*

✳ Most of these animal pioneers were insects or other ARTHROPODS—members of a giant phylum of animals that still dominates life on land. Thanks to their exoskeletons or body cases, they were well placed to counter the threat of drying out.

bat

VERTEBRATE MASTERY

✱ Vertebrates emerged onto land about 30 million years later. Unlike invertebrates, they took longer to establish a hold. The first semiterrestrial vertebrates were the early amphibians—animals like ERYOPS, which was about 6 feet long. Eryops evolved from fleshy-finned fish, but had stubby limbs and was able to breathe air.

✱ By any standards, Eryops was a successful predator, but it was still restricted to swamps and shores. **Full mastery of land had to wait until the evolution of the first reptiles, about 300**

land dwellers
need to avoid
dehydration

million years ago. Unlike amphibians, they had waterproof skins and waterproof eggs, allowing them to spread far inland.

✱ Reptiles proved to be tremendously successful and dominated life on land until the Cretaceous mass extinction (see page 26), 65 million years ago. A catastrophic event for reptiles, it allowed mammals to take center stage.

FIRST FLIGHT

Flight has evolved in four separate groups of living things: insects, pterosaurs (flying reptiles), birds, and bats. The smallest flying insects have a wingspan of a few millimeters, whereas the largest pterosaur— QUETZALCOATLUS— boasted a wingspan of up to 40 feet, making it the biggest flying animal ever to have existed.

THE AMNIOTIC EGG

Reptiles were the first vertebrates to reproduce with the help of amniotic eggs. These eggs contain a membrane called the AMNION, which forms a fluid-filled bag around the embryo. This amniotic fluid allows the embryo to develop in a watery environment, even though the egg is laid on dry land.

I'm going back to my roots

PLANT EVOLUTION

* Plants almost certainly evolved from a group of green algae called the chlorophytes, which abound in fresh water. For organisms like these, life on damp mud poses few major problems, and it is easy to imagine how their predecessors became established on low-lying land. Despite their tiny size, ancient chlorophytes were the ancestors of all land plants, from the tiny daisy to the American giant redwood.

VASCULAR PLANTS

* Some of these early plants developed a waxy surface, or cuticle, that helped stop them from drying out—but by late Silurian times, about 400 million years ago, plants had evolved better ways to tackle their water problem. By developing microscopic internal pipelines, or <u>VASCULAR TISSUE</u>, they were able to move water to where it was needed. <u>*COOKSONIA*</u>, the earliest known vascular plant, was a giant of its time, with branching upright stems towering 2 inches high.

* *Vascular systems proved to be a momentous breakthrough. From now on, plants were no longer restricted to damp ground; instead, their roots could seek out moisture far beneath the surface.*

Fungi

At one time fungi were classified as plants, which is why they are often included in botanical books. But, although they are often "rooted" in the ground, it is now accepted that they are no more closely related to plants than to animals. This realization meant fungi had to be reclassified. They were officially promoted to kingdom status in the 1950s—much to some botanists' annoyance.

a fir cone

you are just such a fun guy!

SEED PLANTS

✳ *Cooksonia* and its relatives reproduced by shedding SPORES—dustlike particles that often contain a single cell. Even when the first trees evolved, creating the immense forests of Carboniferous times, most plants still reproduced this way. The exceptions were an up-and-coming group that made SEEDS—complex reproductive structures containing an embryo and a store of food. Seeds are immensely tough and can survive harsh conditions for decades or even centuries.

bunch of carrots

✳ SEED PLANTS diversified with the evolution of conifers—but their major evolutionary push began about 125 million years ago with the arrival of flowering plants. ***Thanks to their complex relationships with animals, which spread their pollen and their seeds, these have since become the dominant forms of plant life.***

ALTERNATING GENERATIONS

In animals, sex cells exist solely to pass on their single set of genes and do not have any independent existence of their own. In plants, things are very different. After they mature, HAPLOID PLANT CELLS (ones with a single set of genes) start to divide and grow, forming an intermediate generation called a GAMETOPHYTE. Gametophytes then produce male and female cells, which fertilize each other to produce the next generation of parent plants, known as SPOROPHYTES. In "primitive" plants such as ferns, gametophytes are free-living. They are often tiny and usually look nothing like their parents.

a silver fir

171

skull of Piltdown
man, *homo dawsoni*

Piltdown man

In 1912 **Charles Dawson**, an English solicitor and amateur archeologist, made what seemed to be *a stupendous find: a collection of fossilized hominid remains in a gravel bed near Piltdown, in southeastern England. The remains included part of a jaw and some teeth, and they fit together to form a humanlike skull that was apparently at least 200,000 years old.* For several decades "PILTDOWN MAN" *caused mounting confusion among paleontologists, because it seemed to be at odds with other discoveries then being made. But in 1948, chemical dating revealed that the fossils were much more recent than Dawson had claimed, and in 1953 the entire find was denounced as a skillful fake.* By then, Dawson was long dead. His exact role in the hoax has never been established.

HUMAN EVOLUTION

✷ Today, nearly 140 years after Thomas Huxley's famous clash with Bishop Wilberforce, the idea that humans have evolved from earlier primates is almost universally accepted. But, although apes are undoubtedly our closest living relatives, establishing exactly how we diverged from them has proved to be a tricky business.

our oldest
relative

WALKING UPRIGHT

✷ Until recently, many paleontologists believed that the split between the lines leading to modern apes and humans occurred about 15 million years ago. *However, the latest molecular dating techniques show that apes and humans parted company about 4–5 million years ago.*

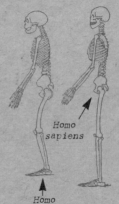

Homo
sapiens

Homo
neanderthalensis

* The earliest fossils that unequivocally belong to the human line have all been discovered in Africa. Known as AUSTRALOPITHECINES, or SOUTHERN APES, they have small canine teeth (unlike true apes) and relatively small brains. There were at least half a dozen types of these hominids, *but all of them shared a new talent: walking upright.*

* One of the most complete australopithecine skeletons was found in 1974 in Ethiopia. Its owner, nicknamed LUCY, belonged to a species called *AUSTRALOPITHECUS AFARENSIS*, and died about 3 million years ago. Lucy was not much bigger than a chimp and had a brain about one-third the size of ours. However, *many paleontologists put her on a direct line leading to modern man.*

THE FIRST HUMANS

* By about 2.5 million years ago, new members of the human family had appeared. *HOMO HABILIS* ("handy-man") was not much bigger than *Australopithecus afarensis*, but had a larger brain and smaller teeth. *For the first time, stone tools start to appear alongside fossil remains.*

* By about 1.5 million years ago, *Homo habilis* had given way to a larger species— HOMO ERECTUS—who was brainier still and could build shelters and use fire. *Unlike* Homo habilis, *this newcomer was an efficient hunter and spread out beyond Africa, throughout Europe and Asia.*

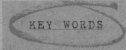

KEY WORDS

HOMINOIDS:
collective name for apes and humans; unlike monkeys, hominoids do not have tails and often hold their bodies upright

HOMINIDS:
members of the family that includes humans and our direct relatives

THE FIRST FOOTPRINTS

The oldest known hominid footprints were discovered at Laetoli, in Tanzania, in 1978. There are three sets of footprints; and *they were preserved by volcanic ash soon after they were made, about 3.6 million years ago.*

THE ORIGIN OF MODERN HUMANS

✱ Our species—*Homo sapiens*—appeared less than 500,000 years ago, slowly replacing our immediate ancestor *Homo erectus*. Humans initially had some archaic features, such as prominent brow ridges, but by about 100,000 years ago they were almost indistinguishable from people alive today.

I wonder when my 500,000th birthday is?

caveman

MITOCHONDRIAL EVE

Data from mitochondrial DNA suggest that the entire human population may be descended from a single female—known, predictably enough, as <u>MITOCHONDRIAL EVE</u>. The mtDNA data also make it likely that many parts of the world have been colonized several times, by waves of people bringing new variants of mtDNA with them.

caveman tools

MERGE OR REPLACE

✱ There are currently two contrasting theories about where anatomically modern humans first appeared.

✱ The single-origin theory—known as the "Out of Africa" hypothesis—holds that *Homo sapiens* evolved in a particular locality in Africa and then spread to other parts of the world. As they spread, they encountered bands of *Homo erectus*, whose distant ancestors had migrated out of Africa many millennia before. The quick-witted *Homo sapiens* prospered, while *Homo erectus* eventually died out.

✱ The opposing view—the multiregional theory—paints a very different picture. It suggests that humans evolved from *Homo erectus* populations in many parts of the world, separately but simultaneously. In taxonomic terms, this would mean that different lines of humans have evolved in parallel, making us members of a "grade" (see page 142).

getting around

THE EVIDENCE

✻ The multiregional theory is backed up by fossil evidence that shows a complete line of human evolution, from archaic to anatomically modern, with no sudden jumps. For its part, the "Out of Africa" hypothesis has some powerful backing in the form of DNA. DNA is not only housed in cell nuclei; a small amount is also found in mitochondria. Unlike nuclear DNA, MITOCHONDRIAL DNA (generally called MtDNA, for short) **is always inherited directly from the mother, making mutations much easier to study.**

hunting

✻ Using the principle of the molecular clock, biochemists have been able to study the mtDNA diversity of the entire human race—and the results show that people living in Africa top the list. This is exactly what would be expected if *Homo sapiens* originally came from that continent, because mutations would have had a long time to build up. **For many paleontologists, this strongly suggests that we all share a recent African past.**

NEANDERTHAL MAN

The disappearance of NEANDERTHAL MAN (see page 92) is one of the greatest puzzles in the evolution of modern humans. Classified variously as a subspecies of *Homo sapiens* or a species in their own right, Neanderthals had bigger brains than our own and were highly efficient hunters. But, after surviving for more than 200,000 years in Europe and northern Asia, they vanished abruptly about 30,000 years ago. Some paleontologists believe that the Neanderthals may have interbred with modern humans, but more likely possibilities are that they were left behind in the competition for resources or killed off by disease.

GENES AND MEMES

In his book The Selfish Gene, *the zoologist Richard Dawkins coined the word* MEME *to describe abstract ideas that can be passed on by cultural exchange.* According to Dawkins, memes constitute a new kind of self-replicating entity, on par with the first self-replicating chemicals that originally gave rise to life. *Dawkins maintains that there is no rule that says evolution has to involve genes: genetic evolution is simply one kind of evolution among many that could be possible.*

CULTURAL EVOLUTION

✱ The most distinctive feature of human beings is that we learn not only through our own experience, but also through the experience of others. This creates cultural exchange—a way of passing on information and ideas that sidesteps the normal processes of evolution.

ANIMAL COPYCATS

✱ *Cultural exchange—also known as* SOCIAL LEARNING—*is rare in the animal world.* One of the most famous cases was reported in the 1960s by biologists studying macaque monkeys on the Japanese island of Koshima. In order to entice the monkeys into the open, the scientists scattered food on a beach. An unusually bright monkey called IMO "invented" a way of cleaning the sand off the food by washing it in a stream. *Within a few weeks the habit caught on, and soon all the monkeys in the group—except the oldest and the very young—were using the technique.* Two years later, Imo made another breakthrough, discovering a way of separating sand from grain. *As before, her discovery was quick to spread.*

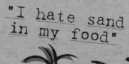

"I hate sand in my food"

monkeys can learn from each other

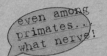

even among primates... what nerve!

STORING INFORMATION

* Despite Imo's remarkable inventiveness, even among primates cultural exchange plays a minor role in life. With humans, however, it looms large. *Thanks to their unusually big brains, our distant ancestors could use cultural exchange to build up and hand on a store of information that was not encoded by genes.* Initially, that information would have been restricted to simple facts, but over hundreds of millennia the information store grew in size and complexity. *In recent centuries, the increase has been exponential, creating the vast inventory of knowledge that we command today.*

EVOLUTION AMONG IDEAS

itsy bitsy spider...

* *Although cultural characteristics are not controlled by genes, they develop in ways that mimic genetic evolution.* They can be copied and handed on, and they also compete in the struggle for survival. In this struggle some variants fare better than others, and the least-favored experience the cultural equivalent of extinction: they are eventually forgotten and abandoned.

Fads and fashions

Changing fashions in clothing and footwear show some striking parallels with biological evolution. For example, most of their features—such as buttons and heels—originally start out as adaptations that have a purely practical role. Over successive generations, some are then exaggerated by sexual selection until eventually, as in the case of spike heels, they become purely ornamental. Clothes also have their own "vestigial organs" (see page 10)—structures like men's lapel buttonholes, which were once used to fasten the lapels beneath the neck.

nursery rhymes are an example of cultural exchange

THE LOOM OF LANGUAGE

✻ At some point in the past, perhaps as long as 2 million years ago, our distant ancestors started to exchange simple ideas by speech. The result, thousands of millennia later, was language —a structured system of communication that has an evolutionary life of its own.

KONICHIWA BONJOUR HOLA BUON GIORNO HELLO GUTEN TAG

A LINK WITH THE PAST

Words that languages share can throw light on daily life at the time when their ancestral protolanguage was in use. For example, many Indo-European languages have similar-sounding words for a number of crops and domesticated animals. These similarities almost certainly mean that these plants and animals were already being farmed when the protolanguage was in use. The names of recently introduced crops and animals do not follow this pattern, because they arrived after the protolanguage became defunct.

False leads

Linguists, like taxonomists, have to be careful not to confuse superficial similarities with ones that have a much longer evolutionary history. Borrowed words can make languages seem more alike than they really are, and so can words that imitate natural sounds.

LANGUAGE FAMILIES

✻ Modern humans probably speak more than 5,000 different languages, although the exact number is difficult to pin down. *Many linguists believe that if modern humans share the same ancestry—as the "Out of Africa" hypothesis suggests (see page 174)—it is reasonable to assume that all these languages have evolved from a common root.*

this goes rather well with cilantro

simple ideas can be exchanged using language

✱ Today's languages can be grouped into about 200 families. English, for example, belongs to the Indo European family, which also includes Celtic, Slavic, and Greek. Similarities between the different Indo-European languages indicate that they all developed from a PROTOLANGUAGE, which was probably spoken somewhere in southern Europe or Central Asia about 7,000 years ago.

LANGUAGE EVOLUTION

✱ *The evolution of language is similar to the evolution of living things, although it often happens much more quickly.* The most frequently used words and phrases—ones denoting being or doing—usually change fastest of all, because they "reproduce" most rapidly. *This explains why the verb "to be" is irregular in almost all languages, while verbs that are used less often behave in more predictable ways.* Mutations—new words or new variants of old ones—appear in languages all the time, and if they prove useful, they quickly spread.

Written relics

Written words—like living things—often contain built-in clues that give hints about their evolutionary past. The English word ghost, for example, did not originally have its unspoken "h"—it was inserted by the 15th-century printer William Caxton, who borrowed it from the Flemish equivalent, *gheest*. The English "h" has remained in place ever since.

Language diversity

In language, as in life, geographical isolation encourages diversity. In the highlands of Papua New Guinea, languages have evolved by the hundred among tribes occupying inaccessible mountain valleys, exactly mirroring what happens to isolated plants and animals. When these isolated languages are exposed to introduced outsiders, the effect can be just a drastic as when new species are brought in: faced with new competition, local languages often have to struggle to survive.

179

CHAPTER 7

I can't keep up with the pace

cultural evolution
moves quickly

DARWINISM AND SOCIETY

* In human history, the pace of cultural evolution has far outstripped biological evolution. Natural selection undoubtedly played a large part in the past of the human race, but experts disagree on how much—if at all—it still influences us today.

"Yellow peril"

Social Darwinism contained a number of contradictions. One of them was that the classes or races that judged themselves to be "superior" were not necessarily the best at passing on their genes—the real test of evolutionary fitness. This created the worrisome possibility that the superior members of society might be swamped by faster-breeding "inferior" stock. In North America, this fear led to the formation of the Asiatic Exclusion League, which successfully campaigned against Chinese immigration.

CLASS AND RACE

* In the late 19th century, *some of Darwin's followers took the view that most aspects of human society could be explained by natural selection*. According to these SOCIAL DARWINIANS, survival of the fittest accounted for the differences between social classes and justified the vast gap in wealth between those at the top and bottom of the pile.

* *The same logic could be used to justify the exploitation of one race by another.* White Europeans assumed that they were the most "highly evolved" branch of the human race, so it was only natural that they should dominate and supervise races that were less technologically advanced. *Decades later, the horrific events of the Nazi era showed where this kind of thinking could lead.*

DIVIDING LINES

* When the pioneer animal-behaviorist **Konrad Lorenz** tried in the 1960s to extrapolate his findings from animals to humans, he met a barrage of criticism. Many people felt that biology was straying into areas where it did not belong.

* In 1975 the American biologist **Edward O. Wilson** published *Sociobiology: the New Synthesis*—a ground-breaking book that analyzed group behavior in terms of its adaptive advantage. It dealt mainly with animals and explored subjects such as altruism, aggression, and sex—but for many readers, its real punch came in the final chapter. *Here, Wilson controversially moved into the human world, exploring the possible evolutionary origins of religion, ethics, territoriality, and war.*

* Wilson conceded that only a small proportion of human behavior is likely to be genetically based, but he helped to create a division that still exists in the scientific world. At one extreme are the full-blown neo-Darwinians, who give genes a prominent role in determining the way we live. *At the other are a large body of anthropologists and social scientists who feel that upbringing, rather than heredity, lies behind the patterns of behavior in human society.*

Hitler

FOR THE GOOD OF THE SPECIES

In the 1880s Charles Darwin's cousin **Francis Galton** suggested that the human species could be artificially improved by selective breeding. Called EUGENICS—literally, "of good stock"—this idea gained momentum after 1900, with the rediscovery of Mendel's work on heredity. Eugenic measures, such as the enforced sterilization of those judged feeble-minded, were carried out in Germany before and during World War II, and in a number of other countries, including some states in the U.S.

Wilson on religion
"It is a reasonable hypothesis that magic and totemism constituted direct adaptations to the environment and preceded formal religion in social evolution"

THE GREAT ESCAPE

✱ Natural selection has helped to make us what we are, but in one area of life we are rapidly escaping its grip. This is old age—a time when genes can do what they like without natural selection being able to control them.

"I'm sure this clock was running fast"

we live longer today

LIFE AND LONGEVITY

Statistics on human life expectancy mask major differences in the fate of individuals. *Fossilized skeletons show that, in preagricultural Europe, average life expectancy was between 25 and 30 years.* Some people managed to survive into their sixties, but the overall average was dragged down by the fact that about 40 percent of children died before their fifth birthday.

"we've escaped our genes"

THE THIRD AGE

✱ *In nature it is very unusual for anything to survive much beyond the time when its reproductive ability comes to an end.* Even the longest-lived trees, which can be over 5,000 years old, are still not botanical senior citizens, in that they still produce seeds every year.

✱ *At one time,* <u>POST-REPRODUCTIVE OLD AGE</u> *was equally rare in humans.* Before farming began, about 10,000 years ago, few people would have been lucky enough to live past their fortieth birthday. *But improvements in*

baby

health and hygiene, particularly during the last hundred years, have changed all that. Today, people in the developed world live long beyond the time when they stop having children, charting a path through a world that has no biological precedent.

GENES AT PLAY

* From an evolutionary perspective, the end of reproduction is a crucial watershed, because it releases genes from the rigors of selection. This may sound like a harmless genetic equivalent of time off, but it actually has some ominous implications. Without the restraining influence of natural selection, genes are free to have any effect, regardless of the results.

* The increasing incidence of degenerative diseases, such as cancer and heart failure, is a direct result of the selective switch-off. In biological terms, a person who has two children and then develops cancer in their fifties is just as "fit" as one who has two children and survives into their nineties. Once you have raised a family, evolution does not care how long you last.

* This makes it the more remarkable that our bodies can last as long as they can. But as human longevity continues to increase, identifying and disabling dangerous late-acting genes becomes ever more important.

Taking off

A steep decline in infant mortality has been the main driving force behind the human population explosion that has occurred during the last hundred years. Human numbers are currently expected to peak at about 11 billion in the year 2075, when the decline in family size finally catches up with the decline in infant deaths.

once you've had children you can drop dead

Final flings

The evolutionary irrelevance of post-reproductive life is vividly shown by animals that breed just once. Mayflies, for example, take to the air without a working digestive system. They mate, lay eggs, and die within the space of a few hours.

on the third day, God created...

evolution contradicts
the biblical
creation story

DOCTRINE AND DOGMA

✳ Belief in a God is not incompatible with a belief that living things evolve. However, the random nature of evolution suggests that if there is divine involvement in life, then it must be much more remote than most of today's religions insist. According to evolutionary theory, creation is a slow and continuous process, not a single event.

THE ANTHROPIC UNIVERSE

Many aspects of the Universe seem to be exactly right for life. For example, nuclear reactions inside stars just happen to form the correct balance of carbon and oxygen atoms for biological molecules to be a possibility. One way of explaining this is to say that it is preordained. Another way, called the ANTHROPIC PRINCIPLE, says that things must be this way because, if they weren't, we wouldn't be here to observe them.

well, it's got nothing to do with me!

MAN DISPOSES

✳ For some biologists, our increasing understanding of the extraordinary complexity of life fully justifies a belief in a "hands-off" deity who has created the conditions for life and then let evolution run its course. *But for many of today's Darwinians, there is no longer any*

Interesting bit

reason to believe in God: life is simply a product of the Universe in which we live.

CREATION SCIENCE

✳ This scientific dogmatism is more than matched by religious zealotry, particularly among fundamentalist Christians. They maintain the belief that all species were specially created, and that none—least of all humans—have departed from their original, divinely crafted design. This viewpoint is known as CREATIONISM, or CREATION SCIENCE.

incremental or instantaneous, it still happened

✳ For most biologists, the position of creationists represents an irrational preference for myths rather than provable facts. *It is claimed that creationism has a scientific basis. Creationism is based on biblical evidence that is said to be above dispute. That, as Darwin would have said, effectively ends the discussion.*

Religion and evolution: two views

"The 'geologic column,' which is cited as physical evidence of evolution occurring in the past, is better explained as the result of a devastating global flood which happened 5,000 years ago, as described in the Bible... There is no reason not to believe that God created the Universe, Earth, plants, animals, and people just as described in the book of Genesis."
Creation Science website

"Today... fresh knowledge has led to the recognition that evolution is more than a hypothesis. It is indeed remarkable that this theory has been progressively accepted by researchers, following a series of discoveries in various fields of knowledge. The convergence, neither sought nor fabricated, of the results of work that was conducted independently, is in itself a significant argument in favor of this theory."
Pope John Paul II, addressing the Pontifical Academy of Sciences in October 1996

CREATION IN COURT

* On July 10, 1925 John T. Scopes, an American high-school teacher, was tried on a charge of violating state law by teaching evolution. The case received worldwide publicity, and it highlighted a campaign by American fundamentalists to squeeze evolution out of the classroom by invoking the law.

yes, I am guilty of teaching evolution

John Scopes was tried for teaching evolution

Scopes II

The American Civil Liberties Union took the state of Arkansas to court in 1981 for passing Act 590—a law that required equal teaching time for evolution and creationism. Billed as SCOPES II, the trial ended with the law being struck down on the grounds that it violated the constitutional separation of church and state. Among the people involved in bringing the action were clergymen who saw Act 590 as a threat to their own religious freedom.

GUILTY AS CHARGED

* The Scopes MONKEY TRIAL took place in Dayton, Tennessee, *one of several states that had banned the teaching of evolution in publicly funded schools.* The judge ruled out any discussion of the underlying issues of evolution and biblical creation, and instead focused entirely on whether or not Scopes had broken the law. He had, and was fined $100. Scopes appealed and was acquitted on the basis that the fine was unjustly large. However, the law that he broke remained in place.

pretty expensive law

no one leaves the classroom until they have learned about evolution

CLASS DISMISSED

✱ In most of Europe, evolution has had a relatively quiet ride in education, based largely on the understanding that it concentrates on the natural world and does not intrude into spiritual affairs. But in the United States, fundamentalism has sharpened the division between science and religion.

✱ William J. Bryan, who led the prosecution against Scopes, declared his intention to drive Darwinism out of schools altogether; and despite Scopes' successful appeal, fundamentalists went on to introduce further anti-evolution bills. *By the mid-1940s, over half of American biology teachers felt they had no option but to avoid evolution altogether.*

✱ *In 1968 the Supreme Court declared that laws banning the teaching of evolution were unconstitutional.* The response by creationists has been to demand equal teaching time for their views. But, as many biologists have pointed out, the balance is not as fair as it seems. Fundamentalist creation science represents just one "traditional" account of creation—an account that is at odds with others from different traditions and different parts of the world.

The Australian Ark

In November 1996 it was the turn of a creationist to be taken to court— this time in Australia. The defendant, Allen Roberts, claimed to have discovered the remains of Noah's ark in Turkey, and sold books and videos describing the remains. Geologist Ian Plimer, who had visited the site, alleged that Roberts had broken Australian trading laws by selling "misleading and deceptive" information. The judge rejected Plimer's case on the grounds that Roberts was not involved in trade or commerce.

ark found in Turkey, read all about it

187

the revolutionary turquoise tomato - it will look great in salads

THE FUTURE OF EVOLUTION

* In nearly 4 billion years, evolution has produced just one form of life that can understand what it means to be alive and to evolve. From the days of our insignificant beginnings, we have reached the point where we will shape the future evolution of other living things and also of ourselves.

THE FINAL THRESHOLD

CLEAN SLATE?

In the future, genetic medicine may be able to "switch off" dangerous genes. Even so, the **background-mutation** rate will mean that hereditary disorders are unlikely simply to vanish. This mutation rate varies with different disorders. For example, intestinal polyposis—a condition that can lead to cancer of the colon—is triggered by a mutant gene that appears spontaneously in about one in 100,000 egg or sperm cells.

* *For most of our time on Earth, humans would have come fairly low down a list of successful evolutionary experiments.* Plenty of other forms of life have been much more widespread than our early ancestors, and many have had a much greater impact on the Earth. But when people started to breed plants and animals, about 10,000 years ago, all that began to change, and *Homo sapiens* took off.

* *Today we are on the threshold of a second revolution—one that may have even greater implications than farming. Through* <u>GENETIC ENGINEERING</u>, *we can now cross the reproductive barriers that normally isolate species from each other. Soon we may be able to create new* <u>GENOTYPES</u> *from an almost unlimited pool*

genetic engineering
has great implications
for farming

*of genes,
something that
evolution has never
been able to do.*

* ***This genetic
pick-and-mix
may turn out to be
immensely valuable for mankind, but it
does carry risks.*** The genotypes in today's
living things have had a vast period of time
to adapt to their environment, whereas
engineered genotypes are completely new.
*No matter what biotech companies say, the
long-term impact on other forms of life is
impossible to assess.*

YOU CHOOSE

* ***Our own species has already loosened
the bonds of natural selection. For us,
genetic engineering may mark the final
break. Once this happens, our
evolution will be entirely in
our own hands.*** Dangerous or
damaging genes will be
sidelined, and useful ones may
become as easy to get hold of
as medicines are today.
*Instead of nature scrutinizing
every variation, "rejecting that
which is bad" and "preserving
and adding up all that is
good," humans will be doing
the selecting.*

* What kind of job will we make of
it? Only the future will tell.

Natural leap

In nature, genes do
occasionally cross the
species barrier. One
example of this
"HORIZONTAL
TRANSMISSION" involves
the common soil-
dwelling bacterium
*Agrobacterium
tumefaciens*, which
attacks plants. If the
bacterium manages to
break into a plant, it
inserts some of its own
DNA into the DNA of
its host, producing
growths called crown
galls. When genetically
modified, this bacterial
Trojan horse can be used
to insert useful genes into
plant cells.

GENETICALLY
ENGINEERED, CHEAP,
AND WHOLESOME

what will be the next
thing in evolution?

189